# Universitext



*Universitext* is a series of textbooks that presents material from a wide variety of mathematical disciplines at master's level and beyond. The books, often well class-tested by their author, may have an informal, personal, even experimental approach to their subject matter. Some of the most successful and established books in the series have evolved through several editions, always following the evolution of teaching curricula, into very polished texts.

Thus as research topics trickle down into graduate-level teaching, first textbooks written for new, cutting-edge courses may make their way into *Universitext*.

For further volumes:
www.springer.com/series/223

# Universitext



*Universitext* is a series of textbooks that presents material from a wide variety of mathematical disciplines at master's level and beyond. The books, often well class-tested by their author, may have an informal, personal, even experimental approach to their subject matter. Some of the most successful and established books in the series have evolved through several editions, always following the evolution of teaching curricula, into very polished texts.

Thus as research topics trickle down into graduate-level teaching, first textbooks written for new, cutting-edge courses may make their way into *Universitext*.

For further volumes:
www.springer.com/series/223

Alexander Prestel • Charles N. Delzell

# Mathematical Logic and Model Theory

## A Brief Introduction

Springer

Prof. Dr. Alexander Prestel
Fachbereich Mathematik und Statistik
Universität Konstanz
Postfach 55 60
78457 Konstanz
Germany
alex.prestel@uni-konstanz.de

Prof. Charles N. Delzell
Department of Mathematics
Louisiana State University
Baton Rouge, Louisiana 70803
USA
Delzell@math.lsu.edu

ISSN 0172-5939 e-ISSN 2191-6675
Universitext
ISBN 978-1-4471-2175-6 e-ISBN 978-1-4471-2176-3
DOI 10.1007/978-1-4471-2176-3
Springer London Dordrecht Heidelberg New York

British Library Cataloguing in Publication Data
A catalogue record for this book is available from the British Library

Library of Congress Control Number: 2011936630

Mathematics Subject Classification (2010): 03-01, 03B10, 03C07, 03C10, 03C98



*Cover design*: VTeX UAB, Lithuania

Printed on acid-free paper

Springer is part of Springer Science+Business Media (www.springer.com)

# Preface to the German Edition

This textbook is based on lecture notes for courses on mathematical logic and model theory that I have been giving in recent years at the University of Konstanz. A goal of the model theory course was to give a thorough and self-contained presentation of the model theoretic aspects of a series of algebraic theories. This was motivated especially by the intent to write a book using the lecture notes that would allow interested mathematicians not trained in this special field to get to know and understand the best known model theoretic results in algebra common at the time. As probably the most interesting example, let us mention only J. Ax and S. Kochen's treatment of "Artin's conjecture" on $p$-adic number fields.

Since the character of model theoretic results and constructions differs quite profoundly from that commonly found in algebra, because of the treatment of formulae as objects, it seems to me indispensable, for a deeper understanding, to become familiar with the problems and methods of mathematical logic. I have therefore preceded the treatment of model theory with an introduction to mathematical logic. From this results a distinct partition of the book into three parts: mathematical logic (Chapter 1), model theory (Chapters 2 and 3) and the model theoretic treatment of several algebraic theories (Chapter 4).

Because of the special goal of this book, I have not aspired to completeness in either the logical or the model theoretic part – this book makes no claim to present either of these two fields as completely as is nowadays common. Instead, I have tried to be as complete and detailed as possible while pursuing the goal stated above. (This explains, among other things, the limited number of exercises.) For further reading I refer the reader to [Shoenfield, 1967] and [Chang–Keisler, 1973–90].

I thank Dr U. Friedrichsdorf and Messrs J. Dix and J. Schmid for a careful reading of the entire text and for numerous suggestions. And I thank Mrs Edda Polte for the careful preparation of this manuscript.

Konstanz, *Alexander Prestel*
March 1986

# Preface to the English Edition

The original purpose of the German edition, written in 1986, was to give an introduction to basic model theory and to the model theoretic treatment of several algebraic theories – in particular, the theory of Henselian valued fields, presenting J. Ax and S. Kochen's result on the "Artin conjecture" on $p$-adic number fields. It was considered indispensable, for a deeper understanding, to base this introduction to model theory on a presentation of formal logic.

The purpose of this new book is to make this kind of introduction available to a larger audience of mathematicians (as German is sometimes considered to be a difficult language). The English version of the 1986 German book is essentially just a translation. It was not our intention to include subsequent developments, since we did not want to add a new book to the many good books that concentrate mainly on model theory, such as [Hodges, 1997], [Marker, 2002] and [Tent–Ziegler, 2011].

Compared with the German version, there are only two major changes: we have considerably increased the number of exercises, and we have added a second appendix, in which we try to explain to the reader the difference between first- and second-order logic.

There are several possible ways to use this book in a course, either for first- or second-year graduate students or for advanced undergraduates. For a course in logic alone (with no model theory), one could cover Chapter 1 and Appendix A (and possibly also Appendix B); for such a course, the student would need little prior knowledge of specific areas of mathematics, but only some mathematical maturity, i.e. some experience in proving theorems. For a course in model theory alone, one can skip Chapter 1 with the exception of the definitions of a formula (Section 1.2), a structure and the satisfaction of a formula by a structure (Section 1.5).[1] For a short course in model theory, Chapters 2 and 3 will suffice (here some familiarity with cardinal and ordinal numbers is required). For a longer course, one could include

[1] In skipping most of Chapter 1 (i.e. formal proofs), the proof of Theorem 2.1.1 will have to rely, instead, on Theorems 2.6.4 and 2.3.3.

Chapter 4 as well; in Sections 4.3–4.6, the student should have some prior acquaintance with valuation theory.

Konstanz and Baton Rouge,
November 2010

*Alexander Prestel*
*Charles N. Delzell*

# Contents

# Introduction

In the middle of the 1960s, several model theoretic arguments and methods of construction captured the attention of the mathematical world. J. Ax and S. Kochen succeeded, in joint work, in achieving a decisive contribution to the "Artin conjecture" on the solvability of homogeneous diophantine equations over $p$-adic number fields. This and other results led to an infiltration of certain model theoretic concepts and methods into algebra. Because of their strangeness, however, very few algebraists could become comfortable with them. This is not so astonishing once one traces the historical development of model theoretic concepts and methods from their origin up to their present-day applications: their applicability to algebra is not the result of a goal-directed development (more precisely, a development directed toward the goal of these applications), but is, rather, a by-product of an investigation directed toward a quite different goal, namely, the goal of getting to grips with the foundations of mathematics. The need to secure these foundations had become urgent after various contradictions in mathematics were discovered at the end of the nineteenth and the beginning of the twentieth centuries.

The attempt to lay a foundation for analysis had led to the development of set theory at the end of the nineteenth century. This development was more and more connected with the use (and, *a fortiori*, the assumption of the existence) of infinite sets that were less and less comprehensible. From a modern-day perspective it is not surprising that contradictions eventually arose from such existence assumptions. Here we may recall Russell's antinomy, in which a contradiction arose from the assumption of the existence of a set whose elements are precisely those sets that do not contain themselves as elements.

One of the most significant proposals for a new foundation of mathematics – besides Brouwer's intuitionism, which in this connection we do not wish to enter into – is David Hilbert's proof theory.

Hilbert proposed taking as the objects of consideration not the mathematical objects themselves, but rather the "statements about these objects". This is to be understood as follows: The objects of our consideration are "sentences" that represent certain mathematical statements, e.g. the statement that some object $A$ with certain properties exists. What is considered, however, is not the object $A$ whose existence


DOI 10.1007/978-1-4471-2176-3_1, 

this "sentence" asserts, but rather the "sentence" itself, written as a finite sequence of letters and symbols in a formal language (assumed to possess only finitely many letters and symbols). The usual "methods of reasoning" in logic are applied to such "sentences". In such an application, however, the content of the "sentences" should not be referred to, but rather only their syntactical structure. Finally, a given set of "sentences" (possibly intended as a set of axioms) should be counted as consistent if no "sentences" that contradict each other (e.g. a "sentence" and its negation) can be deduced from them by means of the above mentioned "methods of reasoning".

If one manages, somehow, to prove the consistency of axioms that imply (among other things) the existence of some infinite totality, one has not thereby actually constructed this totality; rather, one has proved only that its existence can be assumed without harm.

An essential point in this "program" of Hilbert was to declare the permissible means by which the proof of the consistency of the given set of axioms is to be carried out: namely, he required that only finitary reasoning be used, and such reasoning was to be applied to the given axioms, treated as finite strings of symbols, and to any deductions from those axioms (which are themselves finite strings of symbols). The proof of consistency must itself transpire via finitary means; it may not rely upon a "realization" of the axioms, which would possibly again assume the proof of the existence of certain infinite sets that the axioms may stipulate.

In order to capture the entirety of ordinary mathematics in "Hilbert's program", the following would have to be accomplished:

(1) the specification of a (formal) language that allows us to express everything in ordinary mathematics;
(2) the specification of a complete system of universally valid forms of logical reasoning;
(3) the specification of a complete system of mathematical axioms (assumptions); and
(4) a proof of the consistency of the formal system given in (1)–(3).

In this program, all objects must be finitarily or at least effectively constructible, i.e. we require:

for (1): effective constructibility of the alphabet and all syntactical concepts;
for (2): effective constructibility of the system of forms of logical reasoning;
for (3): effective constructibility of the system of mathematical axioms; and
for (4): a proof (of consistency) by "finitary methods".

"Completeness" of the means of logical reasoning means that the addition of further, universally valid forms of logical reasoning would not allow one to deduce more from the given axioms than before. "Completeness" of the axioms means, analogously, that the addition of further axioms likewise would not allow one to deduce more than before (unless this addition causes the set of axioms to become inconsistent). This last implies, in particular, that every sentence $\alpha$ (of the language specified in (1)), or its negation $\neg\alpha$, is deducible from a complete axiom system. Indeed, otherwise one could properly extend this system by the addition of $\alpha$ or $\neg\alpha$.

If the program delineated above could actually be carried out, then a truly ingenious reduction of all of mathematics to "the finite" would be achieved, and thus an unchallengeable foundation for mathematics would be guaranteed. The works of K. Gödel in the 1930s showed, however, that only certain parts of this program are realizable, while others, particularly (4), cannot be realized even in principle.

Task (1) above is relatively simple to carry out. One need only recognize the fact (which is nowadays a commonplace) that all of mathematics can be expressed in the language of set theory. Thus, in order to realize (1), one simply uses the language of set theory. We shall elaborate on this in Section 1.6.

Task (2) was accomplished by Gödel. He showed the completeness of (essentially) the system of logical reasoning used by Hilbert. This is the content of Chapter 1 of our book.

Regarding task (3), Gödel proved that there can be no complete (effectively enumerable) axiom system for the entirety of mathematics, and thus[1] also not for set theory. Gödel's incompleteness proof shows this even for most subfields of mathematics, e.g. arithmetic (the theory of the natural numbers) (the first Gödel incompleteness theorem). This means that for each effectively enumerable system of axioms for arithmetic, there is always a sentence $\alpha$ (in the specified language therefor) such that neither $\alpha$ nor its negation $\neg\alpha$ is deducible from this system. We shall go more deeply into this in Appendix A.

Regarding task (4), Gödel proved that under certain minimal requirements about the formal system, its consistency cannot be proved "by means of this system" (Gödel's second incompleteness theorem). Without going more precisely into it here, let us only mention that "finitary methods", in the usually understood sense, can be carried out in the system of arithmetic and hence naturally in every stronger system, including those that might comprehend the entirety of mathematics. In this sense task (4) cannot be carried out even in principle. However, if we extend our understanding of "finitary methods", then task (4) can indeed be carried out in some cases: in 1936 Gentzen carried this out for the usual Peano axiom system of arithmetic (this system is, however, incomplete by Gödel's first incompleteness theorem).

Regarding our discussion of tasks (3) and (4), the interested reader may refer to more advanced literature, such as [Barwise, 1977], [Hilbert–Bernays, 1934–68], and [Shoenfield, 1967].

Although, as was just mentioned, many fields of mathematics admit no complete, effectively enumerable system of axioms, there are, here and there, fields that do. Two such examples are the algebra of the real numbers and the algebra of the complex numbers. The algebraic theory of the complex numbers can be completely axiomatized by the axioms of an "algebraically closed field of characteristic 0". The

[1] Here we regard the entire field of mathematics as being a subfield of set theory, as we began doing two paragraphs above. During everyday life, however, most mathematicians take the opposite view, that set theory is a subfield of mathematics.

completeness of this axiom system means precisely that every sentence $\alpha$ (in the language of fields), or its negation $\neg\alpha$, is deducible from this axiom system. Since a sentence that is deducible via pure logic from an axiom system naturally holds in every realization (model) of this system, it immediately follows that $\alpha$ holds either in all algebraically closed fields of characteristic 0, or in none of them. This means, in other words, that every sentence $\alpha$ that holds in $\mathbb{C}$ holds in every algebraically closed field of characteristic 0 (and conversely). This is a very simple form of the so-called "Lefschetz principle", a principle for the "transfer" of sentences from one model to another.

Two structures $\mathfrak{A}$ and $\mathfrak{B}$ that cannot, as in the example given just above, be distinguished by sentences $\alpha$ of a given (suitable) formal language $\mathscr{L}$, are called *elementarily equivalent* (with respect to $\mathscr{L}$). Two models of a complete axiom system are therefore always elementarily equivalent (in the language of the axiom system).

The concept of elementary equivalence, and the methods that allow one to construct structures $\mathfrak{B}$ elementarily equivalent to a given structure $\mathfrak{A}$, became central concepts and methods of a subdiscipline of mathematical logic – model theory. This subdiscipline was, over the years, further developed and came into its own more and more. It has its roots, however, for the most part in the controversy over Hilbert's program.

The discussions above should suffice to highlight the central role of the concept of formal sentences for model theory.

In Chapter 1 we shall give an introduction to formal systems, and we shall carry out a proof of Gödel's theorem on the completeness of the system of logical reasoning given in Chapter 1. In the course of this proof we shall, at the same time, become acquainted with the first methods for the construction of elementarily equivalent structures. In Chapters 2 and 3 we shall introduce further typical model theoretic construction methods and concepts, mainly with the goal of proving (in Chapter 4) the completeness of a series of (algebraic) axiom systems. Besides the transfer principles obtained thereby (as indicated above), yet a further result follows from the completeness of an (effectively enumerable) axiom system, namely – as we shall work out in Appendix A – the decidability of the corresponding theory; i.e. it gives the existence of an algorithm which, to each sentence $\alpha$ (in the language of the axiom system considered), decides, in finitely many steps, whether or not $\alpha$ can be deduced from the axiom system.

# Chapter 1
# First-Order Logic

In this chapter we introduce a calculus of logical deduction, called first-order logic, that makes it possible to formalize mathematical proofs. The main theorem about this calculus that we shall prove is Gödel's completeness theorem (1.5.2), which asserts that the unprovability of a sentence must be due to the existence of a counterexample. From the finitary character of a formalized proof we then immediately obtain the Finiteness Theorem (1.5.6), which is fundamental for model theory, and which asserts that an axiom system possesses a model provided that every finite subsystem of it possesses a model.

In (1.6) we shall axiomatize a series of mathematical (in particular, algebraic) theories. In order to show the extent of first-order logic, we shall also give within this framework the Zermelo–Fraenkel axiom system for set theory, a theory that allows us to represent all of ordinary mathematics in it.

## 1.1 Analysis of Mathematical Proofs

In this section we try, by means of an example, to come closer to an answer to the question, "What is a mathematical proof?". For the example to be considered, we assume that we find ourselves in an undergraduate mathematics course in which the field of all real numbers is being introduced axiomatically. Let us further assume that the field properties have already been covered and the order properties are just now being introduced by the following axioms:

(0) $\leq$ is a partial order;

(1) for all $x, y$, either $x \leq y$ or $y \leq x$;

(2) for all $x, y$ with $x \leq y$, we have $x + z \leq y + z$ for all $z$; and

(3) if $0 \leq x$ and $0 \leq y$, then also $0 \leq x \cdot y$.

Then we want to give a proof for the following

*Claim*: $0 \leq x \cdot x$ for all $x$.

A. Prestel, C.N. Delzell, *Mathematical Logic and Model Theory*, Universitext,
DOI 10.1007/978-1-4471-2176-3_2, © Springer-Verlag London Limited 2011

A proof of this could look something like the following:

*Proof*: 1. From (1) we obtain $0 \leq x$ or $x \leq 0$.
2. If $0 \leq x$, then (3) gives $0 \leq x \cdot x$.
3. If, however, $x \leq 0$, then from (2) follows $0 \leq -x$ (where we set $z = -x$).
4. Now (3) again gives $0 \leq (-x) \cdot (-x) = x \cdot x$.
5. Therefore $0 \leq x \cdot x$ holds for all $x$. □

In view of this example of a proof, several remarks are now in order with regard to an exact definition of the concept of "mathematical proof".

*Remark 1.1.1* (on the level of detail in a proof). The level of detail of a proof is as a rule geared toward the background of those for whom the proof is intended. In our example, this was the background of undergraduate mathematics students. For experts, a proof at this level of detail would not be necessary – usually a proof in such a case would consist of the single word "trivial". For nonmathematicians, on the other hand, the above proof might be too short, hence hard to understand. A nonmathematician might not be able to follow it, since certain intermediate steps that are clear to the mathematician are simply omitted, or certain conventions are used that only mathematicians are familiar with. For example, the mathematician writes $0 \leq (-x) \cdot (-x) = x \cdot x$, and actually means the expression:

$$0 \leq (-x) \cdot (-x) \text{ and } (-x) \cdot (-x) = x \cdot x \text{ imply } 0 \leq x \cdot x.$$

It should be clear that for an exact definition of proof, we must strive for the greatest possible fullness of detail, so that the question of whether a given sequence of sentences is a proof is checkable by anyone who knows this definition. Moreover, it should even be possible for a suitably programmed computer to make this determination.

*Remark 1.1.2* (on the choice of a formal language). The *language* used to write a proof out is, as a rule, likewise chosen according to the intended audience. In mathematics it is common to care less about good linguistic style, and much more about unique readability. The example of a proof above can well be taken as typical. From the standpoint of unique readability, however, let us attempt some improvements. Thus, the words "also" (in Axiom 3) or "however" (in line 3 of the proof) can be viewed as purely ornamental. They possess no additional informational content. On the contrary, such ornamental words often cause ambiguities. In the above proof one could also complain that sometimes a generalization "for all $x$" is missing from the beginning of a sentence, and sometimes it appears at the end of such a sentence. This especially can easily lead to ambiguities. In order to be able to give an exact definition for the concept of proof, it is therefore indispensable to agree once and for all upon linguistic conventions that guarantee unique readability.

*Remark 1.1.3* (on the layout of a proof). Normally a proof consists of a finite sequence of statements. Often additional hints are given, as, for example, in line 4 of the above proof. An exact definition of proof should, however, make such things superfluous. Such hints should serve only to promote readability, and should have no influence on whether a given sequence of sentences is a proof or not. As to the utilization of the given space for writing out the proof, it should also be immaterial whether the sequence of these sentences is arranged in a series within one line, or (as in the example above) there is only one sentence per line. For the sake of readability, we shall stick to the latter form.

Considering the criticisms in Remark 1.1.2 above, and using symbolism that is widespread in mathematics (which we shall make precise in the next section), we shall now repeat the above proof. First, however, we want to "formalize" those axioms that occur in the proof:

(1) $\forall xy\,(x \leq y \vee y \leq x)$
(2) $\forall xyz\,(x \leq y \rightarrow x+z \leq y+z)$
(3) $\forall xy\,(0 \leq x \wedge 0 \leq y \rightarrow 0 \leq x \cdot y)$

Now to the claim and the proof:

*Claim*: $\forall x\ 0 \leq x \cdot x$

*Proof*:
1. $(1) \rightarrow (0 \leq x \vee x \leq 0)$
2. $0 \leq x \wedge (3) \rightarrow 0 \leq x \cdot x$
3. $x \leq 0 \wedge (2) \rightarrow 0 \leq -x$
4. $0 \leq -x \wedge (3) \rightarrow 0 \leq (-x) \cdot (-x) = x \cdot x$
5. $\forall x\ \ 0 \leq x \cdot x$

□

In order to take the criticism in Remark 1.1.1 into account somewhat, we could formulate the proof in more detail – say, as follows:

1. $(1) \rightarrow (0 \leq x \vee x \leq 0)$
2. $0 \leq x \wedge (3) \rightarrow 0 \leq x \cdot x$
3. $0 \leq x \rightarrow 0 \leq x \cdot x$
4. $x \leq 0 \wedge (2) \rightarrow x+(-x) \leq 0+(-x)$
5. $x+(-x) \leq 0+(-x) \wedge x+(-x) = 0 \rightarrow 0 \leq 0+(-x)$
6. $0 \leq 0+(-x) \wedge 0+(-x) = -x \rightarrow 0 \leq -x$
7. $0 \leq -x \wedge (3) \rightarrow 0 \leq (-x) \cdot (-x)$
8. $0 \leq (-x) \cdot (-x) \wedge (-x) \cdot (-x) = x \cdot x \rightarrow 0 \leq x \cdot x$
9. $x \leq 0 \rightarrow 0 \leq x \cdot x$
10. $(0 \leq x \vee x \leq 0) \rightarrow 0 \leq x \cdot x$
11. $\forall x\ 0 \leq x \cdot x$

□

Now we can discuss several typical characteristics of our by now already somewhat formalized proof.

A proof is a sequence of expressions, each of which either contains a universally valid, logical fact, or follows purely logically (with the help of the axioms) from earlier sentences in the proof. Thus, the first line contains a universally valid fact. Namely, it has (if we suppress the variable $x$ for a moment) the form

$$\forall y\, \varphi(y) \to \varphi(0),$$

where $\varphi(y)$ is an expression that, in our case, speaks about arbitrary elements $y$ of the real number field. Likewise, lines 2, 4, 5, 6, 7 and 8 represent universally valid implications. Using the axioms (1)–(3) and also the identities $x + (-x) = 0$, $0 + (-x) = -x$ and $(-x) \cdot (-x) = x \cdot x$ (which are also to be used as axioms), lines 3 and 9 result from previous statements by purely logical deductions. Thus, for example, we obtain line 3 from the rule of inference that says: if we have already proved $(\alpha \wedge \beta) \to \gamma$ and, in addition, $\beta$, then we have also thereby proved $\alpha \to \gamma$. In our case, $\beta$ is axiom (3), which, as an axiom, needs no proof, or, in other words, can be assumed to have been proved. We obtain line 10 by applying to lines 3 and 9 the rule of inference: from $\alpha \to \beta$ and $\gamma \to \beta$ follows $(\alpha \vee \gamma) \to \beta$. From lines 1 and 10 we actually obtain, at first, only $0 \leq x \cdot x$. However, since this has been proved for a "fixed but arbitrary" $x$, we deduce $\forall x\ 0 \leq x \cdot x$. Thus, line 11 is likewise a universally valid logical inference.

In the next two sections we want, first, to fix the linguistic framework exactly, and, second, to give an exact definition of proof. Since we shall later have very much to say about formulae and proofs, and must often use induction to prove metatheorems about them, it behooves us to proceed very economically in our definitions of formula and proof. Therefore, we shall not take a large number of rules of inference as a basis, but rather try to get by with a minimum. This has a consequence that gap-free (formal) proofs become very long. Thus the above proof, for example, would, in gap-free form, swell to about 50 lines. However, once we give an exact definition of proof, we shall agree to relax that definition so as to allow the use, in proofs, of so-called "derived" rules of inference. These methods correspond exactly with mathematical practice: in new proofs one refers back, possibly, to already known proofs, without having to repeat them. All that is important is that all gaps could, if necessary, be filled (at least theoretically!).

## 1.2 Construction of Formal Languages

For the definition of proof, it is necessary to describe the underlying formal language more precisely; this is the goal of the present section. The objects of our consideration will be an alphabet, and the words and statements formed therefrom. The formal language itself therefore becomes an object of our investigation. On the other hand, we use informal (mathematical) colloquial language (which we might describe as "mathematical English") to formulate everything that we establish in this investigation of the formal language. This is necessary in order for us to communicate these stipulations and results to the reader. Therefore we must deal with two languages, one being the object of our considerations (which we therefore call the *object language*, or, in other contexts, the *formal language*), and the other being the language in which we talk *about* the object language (we call this second one the *metalanguage*).

The metalanguage will always be the mathematical colloquial language, in which, for example, we occasionally use common abbreviations (such as "iff" for "if and only if"). In the metalanguage, we shall also use the set theoretical conceptual apparatus, as is usual in mathematics. And especially in the second part of this book (the model theoretic part, Chapters 2 and 3), we shall reason in set theory. When the considerations make it necessary, it is, however, also possible to return to the "finitist standpoint", in which one speaks only of finite sequences (or "strings") of symbols (built up from the alphabet of the object language) or of finite sequences of such strings of symbols.

The object language will depend on the subject being considered at the time. For example, if we want to talk about the consistency of mathematics, then we adopt the finitist standpoint and therefore require that the alphabet of the object language considered be finite. If, however, we adopt the model theoretic standpoint, then the alphabet may be an arbitrary set.

Before we come to the definitions, we give yet another hint, this time about a fundamental difficulty. The issues that we pursue here are not common in mathematics. Ordinarily one utilizes only one language: the language in which one communicates something, such as a proof. For a mathematician, writing out a sentence is usually tantamount to claiming that that sentence is true. Thinking about the real numbers, for example, one might write (using the usual abbreviations)

$$\forall x \, \exists y \; x < y,$$

rather than

$$\text{the statement} \quad \forall x \, \exists y \; x < y \quad \text{holds.}$$

But if we want to speak *about* a language, then we must necessarily distinguish between the symbol-sequence $\forall x \, \exists y \, x < y$ and its possible meaning. This and the next section will deal only with *syntactical* questions, i.e. questions such as whether a string of symbols is correctly formed with reference to certain rules of formation.

The *alphabet* of the object language that we consider consists of the following fundamental symbols:

| | | |
|---|---|---|
| logical symbols: | $\neg$ (not) $\wedge$ (and) $\forall$ (for all) $\doteq$ (equals) | |
| variables: | $v_0 \quad v_1 \quad v_2 \quad \ldots \quad v_n \quad \ldots \quad (n \in \mathbb{N} := \{0, 1, 2, \ldots\})$ | |
| relation symbols: | $R_i \quad (\text{for } i \in I)$ | (1.2.0.1) |
| function symbols: | $f_j \quad (\text{for } j \in J)$ | (1.2.0.2) |
| constant symbols: | $c_k \quad (\text{for } k \in K)$ | (1.2.0.3) |
| punctuation: | , ) ( | |

Here $I$, $J$ and $K$ are arbitrary index sets, which may even be empty. If we wish to adopt the finitist standpoint, we can generate the infinitely many variables $v_n$ ($n \in \mathbb{N}$) by means of finitely many basic symbols, say, $v$ and $'$: then instead of the symbol $v_n$, we would write

$$v\underbrace{''''\cdots'}_{n \text{ times}}.$$

One could rewrite the relation, function and constant symbols in an analogous way, in which case the index sets $I, J, K$ would naturally be at most countable.

From these basic symbols we now want to construct certain strings of symbols, which we call terms. Terms will, via a semantical interpretation given later, designate *things*; they are, therefore, possible names. If one keeps this in mind, the following definition of *terms* becomes understandable:

(a) All variables $v_n$ and all constant symbols $c_k$ are terms.
(b) If $t_1,\ldots,t_{\mu(j)}$ are terms, then so is $f_j(t_1,\ldots,t_{\mu(j)})$.
(c) No other strings of symbols are terms.

Here $\mu$ is a function that, to each $j \in J$, assigns the "arity" (= number of arguments) $\mu(j)$ of the function symbol $f_j$; thus, $\mu(j) \geq 1$.

Then Tm, the *set of all terms*, is the smallest set of strings of symbols that contains all $v_n$ and $c_k$ and that, for each $j \in J$, contains $f_j(t_1,\ldots,t_{\mu(j)})$ whenever it contains $t_1,\ldots,t_{\mu(j)}$.

By convention, we may sometimes write $t_1\, f_j\, t_2$ instead of the official $f_j(t_1,t_2)$, in case $\mu(j) = 2$; for example, $t_1 + t_2$ instead of $+(t_1,t_2)$, if $+$ is a binary function symbol.

Next, we construct *formulae*:

(a) If $t_1$ and $t_2$ are terms, then $t_1 \doteq t_2$ is a formula.
(b) If $t_1,\ldots,t_{\lambda(i)}$ are terms, then $R_i(t_1,\ldots,t_{\lambda(i)})$ is a formula.
(c) If $\varphi$ and $\psi$ are formulae and $v$ is a variable, then $\neg\varphi$ and $(\varphi \wedge \psi)$ and $\forall v\, \varphi$ are formulae.
(d) No other strings of symbols are formulae.

Here $\lambda$ is a function that, to each $i \in I$, assigns the "arity" $\lambda(i)$ of the relation symbol $R_i$; again, $\lambda(i) \geq 1$.

Then Fml, the *set of all formulae*, is the smallest set of strings of symbols that contains all strings of the form $t_1 \doteq t_2$ and $R_i(t_1,\ldots,t_{\lambda(i)})$ (these are also called the *atomic* formulae), and that contains $\neg\varphi$ and $(\varphi \wedge \psi)$ as well as $\forall v\, \varphi$ whenever it contains $\varphi$ and $\psi$.

From now on, the notations

| | | |
|---|---|---|
| $t_1, t_2, \ldots$ | | will denote *terms*, |
| $\varphi, \psi, \rho, \tau, \alpha, \beta, \gamma$ | (possibly with subscripts) | will denote *formulae*, and |
| $u, v, w, x, y, z$ | (possibly with subscripts) | will denote *variables*. |

We further employ the following *abbreviations*:

$$\begin{array}{llll} (\varphi \vee \psi) & \text{stands for} & \neg(\neg\varphi \wedge \neg\psi) & \text{(or)} \\ (\varphi \to \psi) & \text{stands for} & \neg(\ \varphi \wedge \neg\psi) & \text{(implies)} \\ (\varphi \leftrightarrow \psi) & \text{stands for} & (\neg(\varphi \wedge \neg\psi) \wedge \neg(\psi \wedge \neg\varphi)) & \text{(equivalent)} \\ \exists v\, \varphi & \text{stands for} & \neg\forall v \neg\varphi & \text{(there exists)} \end{array} \tag{1.2.0.4}$$

And we adopt the following *conventions*, which are customary:

1. $\vee$ and $\wedge$ bind more strongly than $\to$ and $\leftrightarrow$;
2. $\neg$ binds more strongly than $\vee$ and $\wedge$;
3. $t_1 \neq t_2$ stands for $\neg\, t_1 \doteq t_2$; (1.2.0.5)
4. $t_1\, R_i\, t_2$ often stands for $R_i(t_1, t_2)$, in case $\lambda(i) = 2$;
5. $\forall u, v, w, \ldots$ stands for $\forall u \forall v \forall w \ldots$;
6. $\exists x, y, \ldots$ stands for $\exists x \exists y \ldots$;
7. $(\varphi_1 \wedge \varphi_2 \wedge \varphi_3)$ stands for $((\varphi_1 \wedge \varphi_2) \wedge \varphi_3)$ (1.2.0.6)
   (i.e. we group left parentheses together);
8. $(\psi_1 \vee \psi_2 \vee \psi_3 \vee \psi_4)$ stands for $(((\psi_1 \vee \psi_2) \vee \psi_3) \vee \psi_4)$; and (1.2.0.7)
9. we drop outside parentheses when this can lead to no ambiguity.

Thus, according to these conventions, the string of symbols

$$\forall x, y\, (\neg\varphi \wedge \psi \ \to\ \alpha \vee \beta \vee \gamma)$$

stands for

$$\forall x\, \forall y\, ((\neg\varphi \wedge \psi) \ \to\ ((\alpha \vee \beta) \vee \gamma)).$$

Now we wish to enter into the role of variables in formulae. As an example, we consider a formal language (object language) with a relation symbol and a constant symbol. Thus $I = \{0\}$, $J = \emptyset$ and $K = \{0\}$, say. For $R_0$ we write $<$, and for $c_0$ we write 0, for short. Let $\varphi$ denote the formula

$$\exists v_0\, (0 < v_0 \wedge v_0 < v_1) \wedge \forall v_0\, (v_0 < 0 \ \to\ v_0 < v_1).$$

If we think of the usual ordering on the real numbers, then we see that the variables $v_0$ and $v_1$ play different roles in $\varphi$. First, it makes little sense to ask whether $\varphi$ is true in the real numbers. This would begin to make sense only if for $v_1$ we think of a particular real number. Obviously $\varphi$ is true if we think of a positive real number for $v_1$. Consider the case where we think of $v_1$ as 1; then $\varphi$ remains true if we replace $v_0$ by, say, $v_{13}$. The "truth value" of $\varphi$ does not change if we replace the $v_0$s in the first part of the formula by $v_{13}$ and the $v_0$s in the second part by $v_{17}$. This is so not only in the case where $v_1$ is 1, but also in every case. On the other hand, we may not replace the two occurrences of $v_1$ in $\varphi$ by two distinct variables; this would alter the "sense" of $\varphi$ in an essential way. This distinction involving the occurrence of a variable in a formula is captured formally by the following definitions.

In the recursive construction of a "for all" (or "universal") formula $\forall v\, \varphi$, we refer to the subformula $\varphi$ as the *scope* (or *effective range*) of the "quantifier" $\forall v$. We call an occurrence of a variable $v$ in a formula $\psi$ *bound* if this occurrence lies within the scope of a quantifier $\forall v$ used in the construction of $\psi$. Every other occurrence[1] of the variable $v$ in the formula $\psi$ is called *free*. We denote by $\mathrm{Fr}(\psi)$ the set of variables that possess at least one free occurrence in $\psi$. The following equations are easily checked:

$$\begin{aligned}\mathrm{Fr}(\psi) &= \{v \mid v \text{ occurs in } \psi\}, \text{ if } \psi \text{ is atomic;}\\ \mathrm{Fr}(\neg\psi) &= \mathrm{Fr}(\psi);\\ \mathrm{Fr}(\varphi \wedge \psi) &= \mathrm{Fr}(\varphi) \cup \mathrm{Fr}(\psi); \text{ and}\\ \mathrm{Fr}(\forall u\, \varphi) &= \mathrm{Fr}(\varphi) \setminus \{u\}.\end{aligned}$$

The elements of $\mathrm{Fr}(\varphi)$ are called the *free variables* of $\varphi$.

For example, in the formula

$$\forall v_0\, (v_0 < 0 \;\rightarrow\; v_0 < v_1) \wedge \exists v_2\, (0 < v_2 \wedge v_2 < v_0),$$

the variable $v_2$ has only bound occurrences, the variable $v_1$ has only free occurrences and the variable $v_0$ occurs both bound (in the first half) and free (in the second half). Note that the scope of $\forall v_0$ is only $(v_0 < 0 \rightarrow v_0 < v_1)$, and not everything after that $\forall v_0$ symbol.

Later we shall need yet another syntactic operation: the *replacement of a variable* $v$ in a string of symbols $\zeta$ by a term $t$. Let

$$\zeta(v/t) \tag{1.2.0.8}$$

denote the string obtained by replacing each free occurrence of $v$ in $\zeta$ by $t$. If a free occurrence of $v$ in the formula $\varphi$ falls within the scope of a quantifier $\forall u$, and if $u$ occurs somewhere in $t$, then after replacement of $v$ in $\varphi$ by $t$, the variable $u$ will obviously fall within the scope of $\forall u$. If this does not happen for any variable $u$ in $t$, then $t$ is called *free for $v$ in $\varphi$*. In other words, $t$ is free for $v$ in $\varphi$ if no free occurrence of $v$ in $\varphi$ lies within the scope of a quantifier $\forall u$ used in the construction of $\varphi$, where $u$ occurs in $t$.

By analogy with the replacement of a variable, we define the *replacement of a constant* $c_k$ in $\zeta$ by a variable $v$ to mean that every occurrence of $c_k$ in $\zeta$ is replaced by $v$. We denote the result of this replacement by

$$\zeta(c_k/v). \tag{1.2.0.9}$$

This, too, is a syntactic operation, i.e. a manipulation of strings of symbols.

---

[1] We do not count the "occurrence" of $v$ in $\forall v$ as a true occurrence.

If the formula $\varphi$ possesses no free variables (i.e. if $\mathrm{Fr}(\varphi) = \emptyset$), then we call $\varphi$ a *sentence*. We write Sent for the set of sentences:

$$\mathrm{Sent} = \{\varphi \in \mathrm{Fml} \mid \mathrm{Fr}(\varphi) = \emptyset\}.$$

The following syntactic operation transforms a given formula $\varphi$ into a sentence: letting $n$ denote the greatest natural number such that $v_n$ occurs free in $\varphi$, we write $\forall\varphi$ for the formula $\forall v_0, v_1, \ldots, v_n\, \varphi$, which we call the *universal closure* of $\varphi$. Obviously, then, $\forall\varphi$ is a sentence.

The concepts of our formal language that we have introduced in this section depend upon three quantities, which we fixed earlier in this section:

$$\begin{aligned} &\text{the "arity" function } \lambda : I \to \mathbb{N}, \\ &\text{the "arity" function } \mu : J \to \mathbb{N}, \text{ and} \\ &\text{the index set } K. \end{aligned} \tag{1.2.0.10}$$

The entire construction of the language depends, therefore, on the triple

$$L = (\lambda, \mu, K). \tag{1.2.0.11}$$

(Observe that the index sets $I$ and $J$ can be recovered as the domains of definition of $\lambda$ and $\mu$.) When we wish to emphasize this dependence on $L$, we write

$$\mathrm{Tm}(L),\ \mathrm{Fml}(L),\ \mathrm{Sent}(L)$$

instead of

$$\mathrm{Tm},\ \mathrm{Fml},\ \mathrm{Sent}.$$

Since all of these concepts are already determined by $L$, we shall often refer to $L$ itself as the "language." By an *extended language* $L'$ of $L$ we mean a triple

$$L' = (\lambda', \mu', K')$$

such that

1. The function $\lambda' : I' \to \mathbb{N}$ extends $\lambda$, i.e. $I \subseteq I'$ and $\lambda'(i) = \lambda(i)$ for all $i \in I$.
2. The function $\mu' : J \to \mathbb{N}$ extends $\mu$.
3. $K \subseteq K'$.

The following inclusions follow immediately from the definitions:

$$\mathrm{Tm}(L) \subseteq \mathrm{Tm}(L'), \quad \mathrm{Fml}(L) \subseteq \mathrm{Fml}(L'), \quad \mathrm{Sent}(L) \subseteq \mathrm{Sent}(L').$$

Observe further that the variables are the same in both languages. We write Vbl for the *set of variables*. We shall write $L \subseteq L'$ to indicate that $L'$ is an extended language of $L$.

In the following chapters we shall often use the following abbreviations: For a finite *conjunction*

$$(\varphi_1 \wedge \cdots \wedge \varphi_n) \quad \text{we write} \quad \bigwedge_{i=1}^{n} \varphi_i,$$

and for a finite *disjunction*

$$(\psi_1 \vee \cdots \vee \psi_m) \quad \text{we write} \quad \bigvee_{j=1}^{m} \psi_j \tag{1.2.0.12}$$

(recall (1.2.0.6) and (1.2.0.7), respectively). If a formula $\varphi$ has the forms

$$\bigwedge_{i=1}^{n} \bigvee_{j=1}^{m_i} \varphi_{ij} \quad \text{or} \quad \bigvee_{i=1}^{n} \bigwedge_{j=1}^{m_i} \varphi_{ij},$$

where each $m_i \geq 1$ and each $\varphi_{ij}$ is an atomic or a negated atomic formula, then $\varphi$ is said to be in *conjunctive normal form* or in *disjunctive normal form*, respectively. A formula $\varphi$ is in *prenex normal form* if $\varphi$ is of the form

$$Q_1 x_1 \cdots Q_n x_n \, \psi,$$

where each $Q_i$ is either the symbol $\forall$ or the symbol $\exists$, and $\psi$ is quantifier-free.

## 1.3 Formal Proofs

Given a (formal) language $L = (\lambda, \mu, K)$, we now want to define the concept of a (formal) proof.

Let $\Sigma$ be a set of formulae: $\Sigma \subseteq \mathrm{Fml}(L)$. In a proof, we shall allow the elements of this set to appear as "axioms" so to speak. A sequence $\varphi_1, \ldots, \varphi_n$ of formulae is called a *proof* (or *deduction*) of $\varphi_n$ from $\Sigma$ if, for each $i \in \{1, 2, \ldots, n\}$:

$\varphi_i$ belongs to $\Sigma$, (1.3.0.1)

or $\varphi_i$ is a *logical axiom*, (1.3.0.2)

or $\varphi_i$ arises from the application of a *logical rule* to members of the sequence with indices $< i$. (1.3.0.3)

The last line, $\varphi_n$, is sometimes called the *end-formula* of the proof.

The concepts "logical axiom" and "logical rule" used in this definition must now be made precise. We subdivide the *logical axioms* into three categories:

tautologies,
quantifier axioms and

equality axioms.

As *logical rules*, we shall allow:

*modus ponens* and
the generalization rule.

We shall now define these various axioms and rules one by one.

In order to be able to define the concept of a tautology precisely, we must first give a brief introduction the *language of sentential logic*. Its alphabet consists of

$$\rightarrow \quad \wedge \quad ) \quad ( \quad A_0 \quad A_1 \ldots A_n \ldots \quad (n \in \mathbb{N}).$$

From this alphabet we construct *sentential forms*:

(a) $A_0, A_1, \ldots$ are sentential forms.
(b) If $\Phi, \Psi$ are sentential forms, then so are $\neg\Phi$ and $(\Phi \wedge \Psi)$.
(c) No other strings of symbols are sentential forms.

The symbols $A_0, A_1, \ldots$ are called *sentential variables*. By a *truth assignment* $\mathscr{H}$ of the variables $A_0, A_1, \ldots$ we mean a function from the set $\{A_0, A_1, \ldots\}$ to the set $\{T, F\}$ of truth values $T$ (= true) and $F$ (= false). Thus, for every $n \in \mathbb{N}$, either $\mathscr{H}(A_n) = T$ or $\mathscr{H}(A_n) = F$. This truth assignment extends canonically from the set of variables to the set of all sentential forms, as follows:

$$\mathscr{H}(\neg\Phi) = -\mathscr{H}(\Phi) \qquad (1.3.0.4)$$

$$\mathscr{H}(\Phi \wedge \Psi) = \mathscr{H}(\Phi) \cap \mathscr{H}(\Psi). \qquad (1.3.0.5)$$

Here $-$ and $\cap$ are operations defined on the set $\{T, F\}$ by the following tables:

| $-$ | $T$ | $F$ |
|---|---|---|
| | $F$ | $T$ |

| $\cap$ | $T$ | $F$ |
|---|---|---|
| $T$ | $T$ | $F$ |
| $F$ | $F$ | $F$ |

A sentential form $\Phi$ is called a *tautological form* if $\Phi$ receives the value $T$ for every truth assignment $\mathscr{H}$. If the sentential form $\Phi$ contains exactly $n$ distinct sentential variables, then for a proof that $\Phi$ is a tautological form, there are exactly $2^n$ cases to consider: for each variable there are just the two values $T$ and $F$ to "plug in". This calculation can be carried out in general according to the schema of the following example. We test the sentential form

$$\neg((A_0 \wedge A_1) \wedge \neg A_0) \qquad (1.3.0.6)$$

by means of the following table:

| $A_0$ | $A_1$ | $(A_0 \wedge A_1)$ | $\neg A_0$ | $(A_0 \wedge A_1) \wedge \neg A_0$ | $\neg((A_0 \wedge A_1) \wedge \neg A_0)$ |
|---|---|---|---|---|---|
| $T$ | $T$ | $T$ | $F$ | $F$ | $T$ |
| $T$ | $F$ | $F$ | $F$ | $F$ | $T$ |
| $F$ | $T$ | $F$ | $T$ | $F$ | $T$ |
| $F$ | $F$ | $F$ | $T$ | $F$ | $T$ |

Here the first line says that for each truth assignment $\mathscr{H}$ with $\mathscr{H}(A_0) = \mathscr{H}(A_1) = T$, the above sentential form receives the value $T$. The subsequent lines are to be read similarly. Since the last column has all $T$s (and no $F$s), we conclude that (1.3.0.6) is, indeed, a tautological form.

Returning to our formal language $L$, an *instance* of a tautological form, or simply a *tautology*, is the formula obtained from a tautological form $\Phi$ by simply replacing each sentential variable in $\Phi$ by a formula of $L$. It should go without saying that different occurrences of the same sentential variable in $\Phi$ must be replaced by the same formula. Thus, if $\varphi, \psi \in \mathrm{Fml}(L)$, then the formula

$$\neg((\varphi \wedge \psi) \wedge \neg \varphi)$$

is an example of a tautology (in view of (1.3.0.6)), and hence of a logical axiom. If we still employ the abbreviations introduced in Section 1.2 (specifically, (1.2.0.1)), then this axiom takes the form

$$(\varphi \wedge \psi) \rightarrow \varphi.$$

It is advisable to utilize such abbreviations whenever one has to check for a tautological form. The following table follows from the definitions:

| $\varphi$ | $\psi$ | $(\varphi \vee \psi)$ | $(\varphi \rightarrow \psi)$ | $(\varphi \leftrightarrow \psi)$ |
|---|---|---|---|---|
| $T$ | $T$ | $T$ | $T$ | $T$ |
| $T$ | $F$ | $T$ | $F$ | $F$ |
| $F$ | $T$ | $T$ | $T$ | $F$ |
| $F$ | $F$ | $F$ | $T$ | $T$ |

With the help of this table, the calculations needed to test for tautological forms are shortened considerably. Another reason to employ such abbreviations is that they enable mathematicians (who are already familiar with most tautologies) to see them in their well-known form.

Next, the *quantifier axioms* are:

(A1) $\forall x \varphi \rightarrow \varphi(x/t)$, in case $t$ is free for $x$ in $\varphi$ (1.3.0.7)

(A2) $\forall x(\varphi \rightarrow \psi) \rightarrow (\varphi \rightarrow \forall x \psi)$, in case $x \notin \mathrm{Fr}(\varphi)$ (1.3.0.8)

Here $\varphi$ and $\psi$ are formulae, $x$ is any variable and $t$ is any term. Actually, (A1) and (A2) each represent infinitely many axioms (just as each tautological form gives rise to infinitely many tautologies).

Next, the *equality axioms* are:

$$
\begin{array}{ll}
\text{(I1)} & x \doteq x \\
\text{(I2)} & x \doteq y \to (x \doteq z \to y \doteq z) \\
\text{(I3)} & x \doteq y \to (R_i(v,\ldots,x,\ldots,u) \to R_i(v,\ldots,y,\ldots,u)) \\
\text{(I4)} & x \doteq y \to \ f_j(v,\ldots,x,\ldots,u) \doteq f_j(v,\ldots,y,\ldots,u)
\end{array}
\tag{1.3.0.9}
$$

Note that (I3) and (I4) are actually *families* of axioms, one for each $i \in I$ and $j \in J$, respectively. The arity of $R_i$ is $\lambda(i)$, and that of $f_j$ is $\mu(j)$ (recall p. 10). In (I3) and (I4), $x$ (which is an arbitrary variable) may be the variable in any position (even the first or last) in the list of variables of $R_i$ or $f_j$, respectively; the variables in the other positions remain unchanged when $x$ gets replaced by $y$.

Having thus completed our description of the logical axioms, we now describe the logical rules. First, *modus ponens* is a logical rule that can be applied to two lines of a proof in the event that one of those two lines has the form $\varphi \to \psi$ and the other has the form $\varphi$, where $\varphi, \psi \in \mathrm{Fml}(L)$. The result of the application is then the formula $\psi$. We display this rule in the following form:

$$
\frac{\begin{array}{c}\varphi \to \psi \\ \varphi\end{array}}{\psi} \quad \text{(MP)}
$$

The line $\varphi_i$ of a proof $\varphi_1, \ldots, \varphi_n$ arises from application of *modus ponens* in case there are indices $j_1, j_2 < i$ such that $\varphi_{j_1}$ has the form $\varphi_{j_2} \to \varphi_i$.

The *generalization rule* allows us to pass from a line of the form $\varphi$ to a line $\forall x\, \varphi$, where $x$ is an arbitrary variable. The line $\varphi_i$ of a proof $\varphi_1, \ldots, \varphi_n$ arises from application of the generalization rule in case there is an index $j < i$ such that $\varphi_i$ has the form $\forall x\, \varphi_j$. We display this rule in the following form:

$$
\frac{\varphi}{\forall x\, \varphi} \quad (\forall)
$$

Here, finally, the definition of a formal proof (from an axiom system $\Sigma$) is concluded. To elucidate this concept, we give a series of examples. The first examples all have the form of *derived rules* (with a single premise); that is, given a proof with a certain end-formula, these rules show how one could, independent of how the end-formula was obtained, append additional lines to the proof so as to obtain a proof of a certain new line.

First we prove the following derived rule:

$$
\frac{\varphi \wedge \psi}{\varphi} \quad (\wedge\, \mathrm{B}_1)
$$

Assume we have a proof of $(\varphi \wedge \psi)$ (from some set $\Sigma$); let this proof be, say,

$$\begin{array}{l} \varphi_1 \\ \vdots \\ \varphi_{n-1} \\ (\varphi \wedge \psi) \quad ; \end{array}$$

then we extend this proof in the following manner:

$$\begin{array}{ll} \varphi_1 & \\ \vdots & \\ \varphi_{n-1} & \\ (\varphi \wedge \psi) & \\ (\varphi \wedge \psi) \rightarrow \varphi & (1.3.0.10) \\ \varphi & (1.3.0.11) \end{array}$$

(1.3.0.10) above is an instance of a tautological form. (1.3.0.11) arises from its two predecessors via an application of *modus ponens*.

By the same argumentation one obtains, in succession, the following derived rules:

$$\frac{(\varphi \wedge \psi)}{\psi} \quad (\wedge \mathrm{B}_2) \qquad \text{from the tautology } (\varphi \wedge \psi) \rightarrow \psi$$

$$\frac{\varphi}{(\varphi \vee \psi)} \quad (\vee \mathrm{B}_1) \qquad \text{from the tautology } \varphi \rightarrow (\varphi \vee \psi)$$

$$\frac{\psi}{(\varphi \vee \psi)} \quad (\vee \mathrm{B}_2) \qquad \text{from the tautology } \psi \rightarrow (\varphi \vee \psi)$$

$$\frac{\varphi \rightarrow \psi}{\neg\psi \rightarrow \neg\varphi} \quad (\mathrm{CP}) \qquad \text{from the tautology } (\varphi \rightarrow \psi) \rightarrow (\neg\psi \rightarrow \neg\varphi)$$

$$\frac{\varphi \leftrightarrow \psi}{\varphi \rightarrow \psi} \quad (\leftrightarrow \mathrm{B}_1) \qquad \text{from the tautology } (\varphi \leftrightarrow \psi) \rightarrow (\varphi \rightarrow \psi)$$

$$\frac{\forall x\, \varphi}{\varphi(x/t)} \quad (\forall \mathrm{B}) \qquad \text{if } t \text{ is free for } x \text{ in } \varphi$$

$$\frac{\forall x\, (\varphi \rightarrow \psi)}{\varphi \rightarrow \forall x\, \psi} \quad (\mathrm{K}\forall) \qquad \text{if } x \notin \mathrm{Fr}(\varphi)$$

The last two derived rules above are obtained using the logical axioms (A1) and (A2), respectively, in just the same way that we obtained $(\wedge\ \mathrm{B}_1)$

The following derived rules each have two premises; i.e. we assume that we have already proved two lines – the premises. First we consider the important rule of “chain implication”:

$$\frac{\varphi \to \psi \qquad \psi \to \sigma}{\varphi \to \sigma} \quad \text{(KS)}$$

This derived rule can be established as follows.

Suppose

$$\begin{array}{ccc} \varphi_1 & & \psi_1 \\ \vdots & \text{and} & \vdots \\ \varphi_{n-1} & & \psi_{m-1} \\ (\varphi \to \psi) & & (\psi \to \sigma) \end{array}$$

are proofs (say, from $\Sigma_1$ and $\Sigma_2$, respectively). Then we obtain the following proof (from $\Sigma_1 \cup \Sigma_2$):

$$\begin{array}{lr} \varphi_1 & \\ \vdots & \\ \varphi_{n-1} & \\ (\varphi \to \psi) & (1.3.0.12) \\ \psi_1 & \\ \vdots & \\ \psi_{m-1} & \\ (\psi \to \sigma) & (1.3.0.13) \\ (\varphi \to \psi) \to ((\psi \to \sigma) \to (\varphi \to \sigma)) & (1.3.0.14) \\ (\psi \to \sigma) \to (\varphi \to \sigma) & (1.3.0.15) \\ (\varphi \to \sigma) & (1.3.0.16) \end{array}$$

Here (1.3.0.14) is a tautology, (1.3.0.15) is obtained by applying *modus ponens* to (1.3.0.12) and (1.3.0.14), and (1.3.0.16) is obtained by applying *modus ponens* to (1.3.0.13) and (1.3.0.15).

In a similar way one obtains the following derived rules (each with two premises):

$$\frac{\varphi \to \psi \qquad \psi \to \varphi}{\varphi \leftrightarrow \psi} \quad (\leftrightarrow) \qquad \text{from the tautology } (\varphi \to \psi) \to ((\psi \to \varphi) \to (\varphi \leftrightarrow \psi))$$

$$\frac{\varphi \qquad \psi}{(\varphi \wedge \psi)} \quad (\wedge) \qquad \text{from the tautology } \varphi \to (\psi \to (\varphi \wedge \psi))$$

$$\frac{\varphi \to \sigma \qquad \psi \to \sigma}{(\varphi \vee \psi) \to \sigma} \quad (\vee) \qquad \text{from the tautology } (\varphi \to \sigma) \to ((\psi \to \sigma) \to ((\varphi \vee \psi) \to \sigma))$$

One could extend the list of derived rules arbitrarily. In fact, this is the method to make proofs more "bearable". Again and again, arguments crop up that one does not want to repeat every time; rather, one incorporates them gradually into the logical system (as derived rules). The more advanced a mathematician is, the more he masters such rules, and the shorter his proofs become.

Before we give a formal proof of our example in Section 1.1, we wish to state, finally, the following derived rules:

$$\frac{t_1 \doteq t_2}{t_2 \doteq t_1}\ (\mathrm{S}) \qquad \frac{\begin{array}{c} t_1 \doteq t_2 \\ t_2 \doteq t_3 \end{array}}{t_1 \doteq t_3}\ (\mathrm{Tr}) \tag{1.3.0.17}$$

$$\frac{t' \doteq t''}{R_i(t_1,\ldots,t',\ldots,t_{\lambda(i)}) \rightarrow R_i(t_1,\ldots,t'',\ldots,t_{\lambda(i)})}\ (\mathrm{R_i}) \tag{1.3.0.18}$$

$$\frac{t' \doteq t''}{f_j(t_1,\ldots,t',\ldots,t_{\mu(j)}) \doteq f_j(t_1,\ldots,t'',\ldots,t_{\mu(j)})}\ (\mathrm{f_j})$$

We leave the proofs of (S), (Tr) and ($\mathrm{f_j}$) to the reader; we now present that of ($\mathrm{R_i}$). We shall carry out the replacement of $t'$ by $t''$ in the first argument of $R_i$. It will be clear that we shall be able to carry out the replacement of every arbitrary argument of $R_i$ by the same method. Thus we assume that we are given a proof of $t' \doteq t''$:

$$\begin{array}{c} \vdots \\ t' \doteq t'' \end{array} \tag{1.3.0.19}$$

We extend this proof by the following lines, where we choose the variables $x, y, u_2, \ldots, u_{\lambda(i)}$ so that they do not appear in any of the terms $t', t'', t_2, \ldots, t_{\lambda(i)}$:

$$\begin{array}{r} x \doteq y \rightarrow (R_i(x,u_2,u_3,\ldots) \rightarrow R_i(y,u_2,u_3,\ldots)) \qquad (1.3.0.20)\\ \forall x\,(x \doteq y \rightarrow (R_i(x,u_2,u_3,\ldots) \rightarrow R_i(y,u_2,u_3,\ldots))) \\ t' \doteq y \rightarrow (R_i(t',u_2,u_3,\ldots) \rightarrow R_i(y,u_2,u_3,\ldots)) \\ \forall y\,(t' \doteq y \rightarrow (R_i(t',u_2,u_3,\ldots) \rightarrow R_i(y,u_2,u_3,\ldots))) \\ t' \doteq t'' \rightarrow (R_i(t',u_2,u_3,\ldots) \rightarrow R_i(t'',u_2,u_3,\ldots)) \\ \forall u_2\,(t' \doteq t'' \rightarrow (R_i(t',u_2,u_3,\ldots) \rightarrow R_i(t'',u_2,u_3,\ldots))) \\ t' \doteq t'' \rightarrow (R_i(t',t_2,u_3,\ldots) \rightarrow R_i(t'',t_2,u_3,\ldots)) \\ \vdots \qquad\qquad \vdots \qquad\qquad \vdots \\ t' \doteq t'' \rightarrow (R_i(t',t_2,t_3,\ldots) \rightarrow R_i(t'',t_2,t_3,\ldots)) \qquad (1.3.0.21)\\ R_i(t',t_2,t_3,\ldots) \rightarrow R_i(t'',t_2,t_3,\ldots) \qquad (1.3.0.22) \end{array}$$

Here (1.3.0.20) is an equality axiom (I3) (1.3.0.9). Thereafter we alternatingly used the rules (∀) and (∀B), until (1.3.0.21). We applied *modus ponens* to this line and (1.3.0.19) to obtain (1.3.0.22).

Now we want to give a formal proof of $\forall x\, 0 \leq x \cdot x$ from the axiom system $\Sigma = \{(1), \ldots, (5)\}$:

(1) $\forall x,y \quad (x \leq y \vee y \leq x)$
(2) $\forall x,y,z\,(x \leq y \rightarrow x+z \leq y+z)$
(3) $\forall x,y \quad (0 \leq x \wedge 0 \leq y \rightarrow 0 \leq x \cdot y)$
(4) $\forall x,y \quad (x+(-x) \doteq 0 \wedge 0+y \doteq y)$
(5) $\forall x \quad (-x)\cdot(-x) \doteq x \cdot x$

These axioms are sentences of a language $L$ whose symbols are specified as follows:

The index set $I$ contains only one element (say, $I = \{0\}$), and $\lambda(0) = 2$; i.e. $R_0$ is a binary relation symbol. For the sake of readability we write $\leq$ for $R_0$. Here one thinks instinctively of the "less-than-or-equal-to" relation between real numbers, which brings with it the temptation to reason semantically. As agreed, however, we wanted to give a purely formal proof whose correctness could be checked even by a computer.

We want to retain for the function symbols the suggestive notation that we have begun to use for the relation symbol. Here $J$ has three elements, say, $J = \{0,1,2\}$, and $\mu(0) = 1$, $\mu(1) = \mu(2) = 2$. For $f_0, f_1, f_2$ we write $-, +, \cdot$, respectively. Furthermore, we make use of the convention that $x+y$ is written for the term $+(x,y)$. Without these agreements, Axiom (2) would take the following form:

(2) $\forall x,y,z\,(R_0(x,y) \rightarrow R_0(f_1(x,z), f_1(y,z)))$.

The index set $K$ is again a singleton – say, $K = \{7\}$. For $c_7$ we write 0, for short.

When we now finally give a formal proof of $\forall x\, 0 \leq x \cdot x$ from $\Sigma$, we shall number the lines, and at the end of each line point out how it arose. Thus, for example, "(MP 3, 29)" in line 30 indicates that this line came about via an application of *modus ponens* to lines 3 and 29.

| | | |
|---|---|---|
| 1. | $\forall x,y\,(x \leq y \vee y \leq x)$ | (Ax(1)) |
| 2. | $\forall y\,(x \leq y \vee y \leq x)$ | (∀B 1) |
| 3. | $(x \leq 0 \vee 0 \leq x)$ | (∀B 2) |
| 4. | $\forall x,y\,(0 \leq x \wedge 0 \leq y \rightarrow 0 \leq x \cdot y)$ | (Ax(3)) |
| 5. | $\forall y\,(0 \leq x \wedge 0 \leq y \rightarrow 0 \leq x \cdot y)$ | (∀*B* 4) |
| 6. | $(0 \leq x \wedge 0 \leq x \rightarrow 0 \leq x \cdot x)$ | (∀*B* 5) |
| 7. | $0 \leq x \rightarrow 0 \leq x \wedge 0 \leq x$ | (Taut.) |
| 8. | $0 \leq x \rightarrow 0 \leq x \cdot x$ | (KS, 6, 7) |
| 9. | $\forall x,y,z\,(x \leq y \rightarrow x+z \leq y+z)$ | (Ax(2)) |
| 10. | $\forall y,z\,(x \leq y \rightarrow x+z \leq y+z)$ | (∀B 9) |
| 11. | $\forall z\,(x \leq 0 \rightarrow x+z \leq 0+z)$ | (∀B 10) |
| 12. | $x \leq 0 \rightarrow x+(-x) \leq 0+(-x)$ | (∀B 11) |

| | | |
|---|---|---|
| 13. | $\forall x, y\ (x+(-x) \doteq 0 \wedge 0+y \doteq y)$ | (Ax(4)) |
| 14. | $\forall y\ (x+(-x) \doteq 0 \wedge 0+y \doteq y)$ | (∀B 13) |
| 15. | $(x+(-x) \doteq 0 \wedge 0+(-x) \doteq -x)$ | (∀B 14) |
| 16. | $x+(-x) \doteq 0$ | ($\wedge B_1$ 15) |
| 17. | $x+(-x) \leq 0+(-x) \rightarrow 0 \leq 0+(-x)$ | ($R_0$ 16) |
| 18. | $x \leq 0 \rightarrow 0 \leq 0+(-x)$ | (KS 12, 17) |
| 19. | $0+(-x) \doteq -x$ | ($\wedge B_2$ 15) |
| 20. | $0 \leq 0+(-x) \rightarrow 0 \leq -x$ | ($R_0$ 19) |
| 21. | $x \leq 0 \rightarrow 0 \leq -x$ | (KS 18, 20) |
| 22. | $\forall x\ (0 \leq x \rightarrow 0 \leq x \cdot x)$ | (∀ 8) |
| 23. | $0 \leq -x \rightarrow 0 \leq (-x) \cdot (-x))$ | (∀B 22) |
| 24. | $\forall x\ (-x) \cdot (-x) \doteq x \cdot x$ | (Ax(5)) |
| 25. | $(-x) \cdot (-x) \doteq x \cdot x$ | (∀B 24) |
| 26. | $0 \leq (-x) \cdot (-x) \rightarrow 0 \leq x \cdot x$ | ($R_0$ 25) |
| 27. | $0 \leq -x \rightarrow 0 \leq x \cdot x$ | (KS 23, 26) |
| 28. | $x \leq 0 \rightarrow 0 \leq x \cdot x$ | (KS 21, 27) |
| 29. | $(x \leq 0 \vee 0 \leq x) \rightarrow 0 \leq x \cdot x$ | (∨ 8, 28) |
| 30. | $0 \leq x \cdot x$ | (MP 3, 29) |
| 31. | $\forall x\ 0 \leq x \cdot x$ | (∀ 30) |

We have thus, finally, transformed the proof given in the usual mathematical style in Section 1.2, into a formal proof. That this transformation has substantially increased the length of the proof is, as already explicitly mentioned earlier, due to the fact that we have presented derived rules only to a limited extent. Our ambition now, however, is not to gain further familiarity with formal proofs or to simplify them step by step until they are, finally, practicable. We want to leave this example as it is. Instead, we now occupy ourselves with the "reach" of such proofs. This will lead us to make claims in our metalanguage about formal proofs, which we then have to prove. The proofs will be carried out in the usual, informal mathematical style. In this section we wish to prove only a couple of small claims and the so-called deduction theorem.

First another definition. Let $\varphi$ be an $L$-formula and $\Sigma$ a set of $L$-formulae. Then we say that $\varphi$ is *provable* (or *derivable*) *from* $\Sigma$, and write

$$\Sigma \vdash \varphi,$$

if there is a proof $\varphi_1, \ldots, \varphi_n$ from $\Sigma$ whose last formula $\varphi_n$ is identical with $\varphi$.

It is clear that if $\Sigma \vdash \varphi$ holds, then for every set $\Sigma'$ of formulae with $\Sigma \subseteq \Sigma'$, $\Sigma' \vdash \varphi$ also holds. In the following claims and their proofs, we become acquainted with further properties of the metalinguistic relation $\vdash$.

**Lemma 1.3.1.** *Let* $\varphi, \psi \in \mathrm{Fml}(L)$, $\Sigma \subseteq \mathrm{Fml}(L)$ *and* $x \in \mathrm{Vbl}$. *Then the following hold:*

*(a)* $\Sigma \vdash \varphi$ *if and only if* $\Sigma \vdash \forall x\,\varphi$
*(b)* $\Sigma \cup \{\psi\} \vdash \varphi$ *if and only if* $\Sigma \cup \{\forall x\,\psi\} \vdash \varphi$.

*Proof*: (a) One reasons from left to right, thusly:

$$\text{If} \quad \begin{array}{c} \vdots \\ \varphi \end{array} \quad \text{is a proof from } \Sigma\text{, then so is} \quad \begin{array}{c} \vdots \\ \varphi \\ \forall x\,\varphi \end{array}.$$

From right to left we use ($\forall$ B):

$$\text{If} \quad \begin{array}{c} \vdots \\ \forall x\,\varphi \end{array} \quad \text{is a proof from } \Sigma\text{, then so is} \quad \begin{array}{c} \vdots \\ \forall x\,\varphi \\ \varphi \end{array}.$$

Observe here that $\varphi(x/x)$ is identical with $\varphi$, and that ($\forall$ B) may be used, since $x$ is naturally free for $x$ in $\phi$.

(b) One reasons from left to right, thusly:

$$\text{If} \quad \begin{array}{c} \vdots \\ \psi \\ \vdots \\ \varphi \end{array} \quad \text{is a proof from } \Sigma \cup \{\psi\}\text{, then} \quad \begin{array}{c} \vdots \\ \forall x\,\psi \\ \psi \\ \vdots \\ \varphi \end{array} \quad \text{is a proof from } \Sigma \cup \{\forall x\,\psi\}.$$

Here the rule ($\forall$ B) is again used. The reasoning from right to left goes as follows:

$$\text{If} \quad \begin{array}{c} \vdots \\ \forall x\,\psi \\ \vdots \\ \varphi \end{array} \quad \text{is a proof from } \Sigma \cup \{\forall x\,\psi\}\text{, then} \quad \begin{array}{c} \vdots \\ \psi \\ \forall x\,\psi \\ \vdots \\ \varphi \end{array} \quad \text{is a proof from } \Sigma \cup \{\psi\}. \qquad \square$$

By repeated use of Lemma 1.3.1(a), we see that the derivability of a formula $\varphi$ from $\Sigma$ is equivalent to the derivability of its universal closure $\forall\,\varphi$ (p. 13) from $\Sigma$. Similarly, repeated use of Lemma 1.3.1(b) permits us to replace all formulae in $\Sigma$ by their universal closures. On the basis of this, we shall often, in the future, limit ourselves to the case where $\Sigma$ is a set of sentences, and $\varphi$ is a sentence.

Now, however, again let $\varphi, \psi \in \mathrm{Fml}(L)$ and $\Sigma \subseteq \mathrm{Fml}(L)$. If

$$\Sigma \vdash (\varphi \to \psi)$$

holds, then one obtains immediately via (MP):

$$\Sigma \cup \{\varphi\} \vdash \psi.$$

Indeed, if

$$\begin{array}{c} \vdots \\ \varphi \to \psi \end{array}$$

is a proof from $\Sigma$, then

$$\begin{array}{c} \vdots \\ \varphi \to \psi \\ \varphi \\ \psi \end{array}$$

is a proof from $\Sigma \cup \{\varphi\}$. Here it is immaterial whether $\varphi$ contains free variables or not. If, however, one knows that $\mathrm{Fr}(\varphi) = \emptyset$, then the above implication can be reversed. We have the following theorem, which is very important for applications:

**Theorem 1.3.2** (Deduction Theorem). *Let* $\Sigma \subseteq \mathrm{Fml}(L)$, $\varphi, \psi \in \mathrm{Fml}(L)$ *and* $\mathrm{Fr}(\varphi) = \emptyset$. *Then from* $\Sigma \cup \{\varphi\} \vdash \psi$ *we obtain* $\Sigma \vdash (\varphi \to \psi)$.

*Proof*: We shall show, by induction on $n$:

if $\begin{array}{c} \varphi_1 \\ \vdots \\ \varphi_n \end{array}$ is a proof from $\Sigma \cup \{\varphi\}$, then the sequence of formulae $\begin{array}{c} \varphi \to \varphi_1 \\ \vdots \\ \varphi \to \varphi_n \end{array}$

can be completed so as to become a proof from $\Sigma$ whose last line remains $\varphi \to \varphi_n$. It is clear that the claim of the Deduction Theorem will follow from this.

*Induction basis step*: Since $n = 1$, we are given a one-line proof from $\Sigma$, consisting of the line $\varphi_1$.

*Case 1*: $\varphi_1$ is a logical axiom or a member of $\Sigma$. In this case,

$$\begin{array}{l} \varphi_1 \\ \varphi_1 \to (\varphi \to \varphi_1) \\ \varphi \to \varphi_1 \end{array}$$

is clearly a proof from $\Sigma$. Here we obtained the last line from $\varphi_1$ and the tautology $\varphi_1 \to (\varphi \to \varphi_1)$ via (MP).

*Case 2*: $\varphi_1$ is identical with $\varphi$. In this case the implication $\varphi \to \varphi_1$ is a tautology, and hence in particular a proof from $\Sigma$.

*Step from n to n + 1*: Assume that

$\begin{array}{c} \varphi_1 \\ \vdots \\ \varphi_n \end{array}$ is a proof from $\Sigma \cup \{\varphi\}$, and the sequence of formulae $\begin{array}{c} \varphi \to \varphi_1 \\ \vdots \\ \varphi \to \varphi_n \end{array}$

has already been completed to a proof from $\Sigma$, with end-formula $\varphi \to \varphi_n$.

*Case 1*: $\varphi_{n+1}$ is a logical axiom or a member of $\Sigma$. In this case we extend the right-hand proof with the lines

$$\varphi_{n+1}$$
$$\varphi_{n+1} \to (\varphi \to \varphi_{n+1})$$
$$\varphi \to \varphi_{n+1},$$

and again obtain a proof from $\Sigma$.

*Case 2*: $\varphi_{n+1}$ is identical with $\varphi$. In this case we simply add the tautology $\varphi \to \varphi_{n+1}$ as the last line.

*Case 3*: $\varphi_{n+1}$ is obtained by (MP). In this case there are $i, j \leq n$ such that the formula $\varphi_j$ has the form $\varphi_i \to \varphi_{n+1}$. But then the lines $\varphi \to \varphi_i$ and $\varphi \to (\varphi_i \to \varphi_{n+1})$ occur in the right-hand proof. We extend the right-hand proof with the following lines:

$$(\varphi \to (\varphi_i \to \varphi_{n+1})) \to (\varphi \to \varphi_i) \to (\varphi \to \varphi_{n+1})) \qquad (1.3.2.1)$$
$$(\varphi \to \varphi_i) \to (\varphi \to \varphi_{n+1}) \qquad (1.3.2.2)$$
$$\varphi \to \varphi_{n+1}. \qquad (1.3.2.3)$$

Here (1.3.2.1) is a tautology, and (1.3.2.2) and (1.3.2.3) are obtained via applications of (MP).

*Case 4*: $\varphi_{n+1}$ is obtained via ($\forall$). In this last case there exists $i \leq n$ such that $\varphi_{n+1}$ has the form $\forall x\, \varphi_i$, where $x$ is a variable. If we extend the right-hand proof with the lines $\forall x\,(\varphi \to \varphi_i)$ and $\varphi \to \varphi_{n+1}$ (in that order), then we shall again obtain a proof of $\varphi \to \varphi_{n+1}$ from $\Sigma$, since these two new lines are justified by the rule ($\forall$) and the derived rule (K $\forall$), respectively. This application of (K $\forall$) is correct, since obviously $x \notin \mathrm{Fr}(\varphi)$ (by the hypothesis that $\mathrm{Fr}(\varphi) = \emptyset$) and $\varphi_{n+1}$ is $\forall x\, \varphi_i$. □

## 1.4 Completeness of First-Order Logic

The preceding sections have shown that it is possible to give a strict, formal definition of the concept of a mathematical proof. It remains to clarify the question of whether this definition really captures what one ordinarily understands by a proof. The unwieldiness that came to light in our earlier example (proving $0 \leq x \cdot x$ from the axioms for ordered fields) can, in principle, be eliminated by the introduction of more and more derived rules. Thus, this unwieldiness is no genuine argument against our formal system. A further possible objection could be that the formal languages we use have inadequate expressive power. But this objection can also be refuted: in Section 1.6 we shall formalize set theory in such a language. Set theory has enough expressive power to express every reasonable mathematical concept.

A completely different objection could be brought against the strength of such formal proofs. It is conceivable that the definition of proof that we gave in the previous section overlooks some valid form of logical reasoning. We wish to show now that this is not the case – i.e. that our concept of proof *completely* comprehends all valid forms of logical reasoning. A heuristic reflection should elucidate this claim.

Let $\Sigma \subseteq \mathrm{Sent}(L)$ and $\varphi \in \mathrm{Sent}(L)$. We ask:

*How might it happen that $\varphi$ is not provable from $\Sigma$ – i.e. $\Sigma \nvdash \varphi$?* (1.4.0.1)

One possibility could be that there is a "counterexample". This should mean that a domain (a mathematical structure – cf. Section 1.5) could exist in which all axioms $\sigma \in \Sigma$ hold, but not $\varphi$. Here we implicitly assume that formal proofs are sound, i.e. that everything that is provable holds wherever the axioms hold. (In Section 1.5 we shall make this precise and prove it.) Another possible reason for the unprovability of $\varphi$ from $\Sigma$ could be that we forgot some method of reasoning when we defined "proof", in which case $\varphi$ could be unprovable even though there is no "counterexample". We shall show that this second case cannot arise:

*Any unprovability rests necessarily on a counterexample.* (1.4.0.2)

We shall prove this claim in Theorem 1.5.2 (Gödel's Completeness Theorem) in the next section, but we shall need several technical preparations, which we would like to carry out in this section. Although we postpone to the next section a precise description of the concept of a structure, and of the definition of the satisfaction of a formula in such a structure, the "counterexample" that we shall construct from the assumption that $\Sigma \nvdash \varphi$ will take on a clear form already by the end of this section.

First we would like to undertake a small reformulation of the hypothesis. For this we call a set $\Sigma \subseteq \mathrm{Sent}(L)$ *consistent* if there is no $L$-sentence $\alpha$ for which both

$$\Sigma \vdash \alpha \quad \text{and} \quad \Sigma \vdash \neg\alpha$$

hold simultaneously. If there is such an $\alpha$, then $\Sigma$ is called *inconsistent*. Then obviously:

$\Sigma$ is inconsistent if and only if one can prove every sentence $\beta$ from $\Sigma$. (1.4.0.3)

Indeed, if $\Sigma$ is inconsistent, then there is a proof of $(\alpha \wedge \neg\alpha)$ from $\Sigma$, by rule $(\wedge)$. We extend this proof by the lines

$$(\alpha \wedge \neg\alpha) \rightarrow \beta \qquad (1.4.0.4)$$
$$\beta. \qquad (1.4.0.5)$$

Here (1.4.0.4) is a tautology, and (1.4.0.5) is obtained with (MP). Using (1.4.0.3) we now show:

**Lemma 1.4.1.** *Let $\Sigma \subseteq \mathrm{Sent}(L)$ and $\varphi \in \mathrm{Sent}(L)$. Then $\Sigma \nvdash \varphi$ if and only if $\Sigma \cup \{\neg\varphi\}$ is consistent.*

*Proof*: We show that $\Sigma \vdash \varphi$ is equivalent to $\Sigma \cup \{\neg\varphi\}$ being inconsistent. First suppose $\Sigma \vdash \varphi$. Then $\Sigma \cup \{\neg\varphi\} \vdash \varphi$ on the one hand, and in any case $\Sigma \cup \{\neg\varphi\} \vdash \neg\varphi$ on the other, whence $\Sigma \cup \{\neg\varphi\}$ is inconsistent.

Now suppose $\Sigma \cup \{\neg\varphi\}$ is inconsistent. Then $\Sigma \cup \{\neg\varphi\} \vdash \varphi$, by (1.4.0.3). Using the Deduction Theorem (1.3.2) we obtain

$$\Sigma \vdash (\neg\varphi \to \varphi).$$

We take some particular proof of $(\neg\varphi \to \varphi)$ from $\Sigma$, and extend it with the lines

$$(\neg\varphi \to \varphi) \to \varphi \tag{1.4.1.1}$$

$$\varphi. \tag{1.4.1.2}$$

Here (1.4.1.1) is a tautology, and (1.4.1.2) is obtained via (MP). Thus $\Sigma \vdash \varphi$ holds. □

Our assumption $\Sigma \nvdash \varphi$ is therefore equivalent to the consistency of the set $\Sigma \cup \{\neg\varphi\}$ of sentences. On the other hand, a "counterexample" to $\Sigma \vdash \varphi$ is exactly a domain in which all $\sigma \in \Sigma \cup \{\neg\varphi\}$ hold. (If $\varphi$ does not hold, then obviously $\neg\varphi$ holds.) In order to produce the "completeness proof" of (1.4.0.2) that we seek, it therefore clearly suffices to do the following:

*to construct, for any consistent set*[2] $\Sigma$ *of sentences,*
*a domain in which all* $\sigma \in \Sigma$ *hold.* (1.4.1.3)

This is what we would like to do now. For this, we shall pursue the following strategy. By means of a systematic, consistent expansion of the set $\Sigma$, we wish to determine the desired domain as far as possible. The steps toward this goal are of a rather technical nature and will be completely motivated and clear only later.

In the *first step* of (1.4.1.3) (which is also the most difficult), we wish to arrive at a stage in which, whenever an existence sentence holds in the domain to be constructed, this can be verified by an example. One such example should be capable of being named by a constant symbol $c_k$ in our language. To achieve this, we shall be forced to extend the given language $L$ by the addition of new constant symbols. We prove the following:

**Theorem 1.4.2.** *Let* $\Sigma \subseteq \mathrm{Sent}(L)$ *be consistent. Then there is a language* $L' \supseteq L$ *with* $I' = I$ *and* $J' = J$*, and there is a consistent set* $\Sigma' \subseteq \mathrm{Sent}(L')$ *with* $\Sigma \subseteq \Sigma'$*, such that to each* $L'$*-existence sentence*[3] $\exists x\,\varphi$*, there exists a* $k \in K'$ *such that*

$$(\exists x\,\varphi \to \varphi(x/c_k)) \text{ is a member of } \Sigma'.$$

For the proof of this theorem we need the following:

**Lemma 1.4.3.** *Let* $L^{(1)} \subseteq L^{(2)}$ *be two languages with* $I^{(1)} = I^{(2)}$, $J^{(1)} = J^{(2)}$ *and* $K^{(2)} = K^{(1)} \cup \{0\}$*, where* $0 \notin K^{(1)}$*. Further, let* $\varphi_1, \dots, \varphi_n$ *be a proof in* $L^{(2)}$ *from* $\Sigma := \{\varphi_1, \dots, \varphi_m\}$*, with* $m \le n$*. Then, if* $y$ *is a variable not occurring in any* $\varphi_i$ $(1 \le i \le n)$*, then* $\varphi_1(c_0/y), \dots, \varphi_n(c_0/y)$ *is a proof in* $L^{(1)}$ *from* $\{\varphi_1(c_0/y), \dots, \varphi_m(c_0/y)\}$.

---

[2] Previously $\Sigma \cup \{\neg\varphi\}$.

[3] Note that this notion differs from that of an "$\exists$-sentence" introduced in Theorem 2.5.4.

*Proof* (of Lemma 1.4.3): We apply induction on the length $n$ of the given formal proof.

*Basis step*: If, in the case $n = 1$, also $m = 1$, then there is nothing to show. If, on the other hand, $m = 0$ (i.e. $\Sigma = \emptyset$), then $\varphi_n$ must be a logical axiom. In the case of an equality axiom, there is again nothing to show, since such an axiom contains no constant symbols. If $\varphi_n$ is an instance of a tautological form, then $\varphi_n(c_0/y)$ is, likewise, clearly an instance of the same tautological form. There remains the case where $\varphi_n$ is a quantifier axiom. So let $\varphi_n$ be of the form (A1) (1.3.0.7):

$$\forall x\, \psi \to \psi(x/t),$$

where $t$ is a term free for $x$ in $\psi$. Now one can easily convince oneself that:

$$\psi(x/t)(c_0/y) \quad \text{is just} \quad \psi(c_0/y)(x/t(c_0/y)).$$

Since $y$ does not occur in $\varphi_n$ by hypothesis, $t(c_0/y)$ is free for $x$ in $\psi(c_0/y)$. Therefore $\varphi_n(c_0/y)$ takes the form of an axiom (A1), namely,

$$\forall x\, \psi(c_0/y) \to \psi(c_0/y)(x/t(c_0/y)).$$

If $\phi_n$ has the form of (A2) (1.3.0.8), one may convince oneself just as easily that $\varphi_n(c_0/y)$ is again an axiom of type (A2).

*Step from $n-1$ to $n$*: Already we know that $\varphi_1(c_0/y), \ldots, \varphi_{n-1}(c_0/y)$ is a proof in $L^{(1)}$ from $\varphi_1(c_0/y), \ldots, \varphi_m(c_0/y)$, where we can assume, without loss of generality, that $m \leq n-1$. Now if $\varphi_n$ is a logical axiom, then, as we saw above, $\varphi_n(c_0/y)$ is again a logical axiom. Two cases remain, in which $\varphi_n$ is obtained by a rule.

*Case 1*: $\varphi_n$ is obtained by (MP). In this case there are $i, j \leq n-1$ such that $\varphi_j$ has the form $(\varphi_i \to \varphi_n)$. Then $\varphi_j(c_0/y)$ has the form $\varphi_i(c_0/y) \to \varphi_n(c_0/y)$. Therefore $\varphi_n(c_0/y)$ is likewise obtained by (MP).

*Case 2*: $\varphi_n$ is obtained via $(\forall)$. In this case there is an $i \leq n-1$ such that $\varphi_n$ has the form $\forall x\, \varphi_i$. Then $\varphi_n(c_0/y)$ has the form $\forall x\, \varphi_i(c_0/y)$. Therefore $\varphi_n(c_0/y)$ is likewise obtained via $(\forall)$.

We observe, finally, that for every $L^{(2)}$-formula $\psi$, the replacement of $c_0$ by a variable leads to an $L^{(1)}$-formula. Thus it is clear that the resulting proof is in $L^{(1)}$.

□

*Proof* (Theorem 1.4.2): We shall obtain the language $L'$ and the set $\Sigma'$ of sentences by a countable process. For each $n \in \mathbb{N}$, we recursively construct a language $L_n$ in the following way: let $L_0$ be the language $L$. For $n \geq 1$, if $L_{n-1}$ has already been constructed, then we obtain $L_n$ by setting $I_n = I_{n-1}$, $J_n = J_{n-1}$ and $K_n = K_{n-1} \cup M_n$. Here $M_n$ is a set disjoint from $K_{n-1}$ such that there is a bijection

$$g_n : M_n \to \{\exists x\, \varphi \mid \exists x\, \varphi \ \text{ is a member of } \mathrm{Sent}(L_{n-1})\} \tag{1.4.3.1}$$

from $M_n$ to the set of all existence sentences in the language $L_{n-1}$. This means nothing more than that we "enumerate" all existence sentences in $L_{n-1}$ with new indices in a one-to-one and onto manner. Sets $M_n$ and bijections $g_n$ of the required kind always exist.

We thereby obtain an ascending chain

$$L_0 \subseteq L_1 \subseteq \cdots \subseteq L_{n-1} \subseteq L_n \subseteq \cdots \tag{1.4.3.2}$$

of languages. Finally we set $L' = \bigcup_{n\in\mathbb{N}} L_n$, i.e. we set $I' = I$, $J' = J$ and $K' = \bigcup_{n\in\mathbb{N}} K_n$. From this one sees immediately that

$$\mathrm{Sent}(L') = \bigcup_{n\in\mathbb{N}} \mathrm{Sent}(L_n)$$

also holds. Therefore if $\exists x\,\varphi$ is an $L'$-sentence, then it lies already in a set $\mathrm{Sent}(L_{n-1})$ for some $n \in \mathbb{N}$.

Suppose

$$\Sigma = \Sigma_0 \subseteq \Sigma_1 \subseteq \cdots \subseteq \Sigma_{n-1} \subseteq \Sigma_n \subseteq \cdots \tag{1.4.3.3}$$

is an ascending chain of sets such that for each $n \in \mathbb{N}$:

$$\begin{aligned} &(1)\ \Sigma_n \subseteq \mathrm{Sent}(L_n), \text{ and} \\ &(2)\ \text{for each } k \in M_n,\ (\exists x\,\varphi \to \varphi(x/c_k)) \text{ is a member of } \Sigma_n, \end{aligned} \tag{1.4.3.4}$$

where $\exists x\,\varphi$ is $g_n(k)$. Once we have such a chain, we shall be able to take $\Sigma'$ to be $\bigcup_{n\in\mathbb{N}} \Sigma_n$. Then it will remain only to check that $\Sigma'$ is consistent.

We obtain a chain (1.4.3.3) with satisfying (1.4.3.4) by setting

$$\Sigma_n := \Sigma_{n-1} \cup \{(\exists x\,\varphi \to \varphi(x/c_k)) \mid k \in M_n,\ g_n(k) \text{ is } \exists x\,\varphi\}. \tag{1.4.3.5}$$

Since $g_n$ is surjective, each existence sentence $\exists x\,\varphi$ in $\mathrm{Sent}(L_{n-1})$ gets counted in (1.4.3.5) – $\exists x\,\varphi$ is, say, $g_n(k)$. Since $\mathrm{Fr}(\varphi) \subseteq \{x\}$, $\varphi(x/c_k)$ is again a sentence (but now in $L_n$, not $L_{n-1}$). Therefore $\Sigma_n \subseteq \mathrm{Sent}(L_n)$. The most important property of $\Sigma_n$ is now its consistency. This results by induction on $n$.

For $n = 0$, the consistency of $\Sigma_0$ is the hypothesis. For $n \geq 1$, suppose that $\Sigma_{n-1}$ is consistent, but not $\Sigma_n$. Then we would obtain an $\alpha$ in $\mathrm{Sent}(L_n)$ with

$$\Sigma_n \vdash (\alpha \wedge \neg\alpha).$$

Since any single proof from $\Sigma_n$ can be traced back to only finitely many axioms of $\Sigma_n$, $(\alpha \wedge \neg\alpha)$ would already be provable from $\Sigma_{n-1}$, together with finitely many sentences

$$(\exists x_1\,\varphi_1 \to \varphi_1(x_1/c_{k_1})),\ \ldots,\ (\exists x_r\,\varphi_r \to \varphi_r(x_r/c_{k_r})), \tag{1.4.3.6}$$

where $g_n(k_i)$ is $\exists x_i\,\varphi_i$, which is a member of $\mathrm{Sent}(L_{n-1})$ for $1 \leq i \leq r$. We briefly write $\sigma_1, \ldots, \sigma_r$ for the $r$ sentences in (1.4.3.6). Then we have

$$\Sigma_{n-1} \cup \{\sigma_1, \ldots, \sigma_r\} \vdash (\alpha \wedge \neg\alpha).$$

By means of a possible (still finite) expansion of the set $\{\sigma_1,\dots,\sigma_r\}$ we can ensure, in addition, that there is a proof of $(\alpha\wedge\neg\alpha)$ already in the sublanguage $L^{(2)}$ of $L_n$ defined by

$$I^{(2)}=I_n,\quad J^{(2)}=J_n,\quad \text{and}\quad K^{(2)}=K_{n-1}\cup\{g_n^{-1}(\exists x_1\,\varphi_1),\dots,g_n^{-1}(\exists x_r\,\varphi_r)\}.$$

Note that $r\geq 1$, since otherwise $\Sigma_{n-1}$ would be inconsistent. Thus, by the Deduction Theorem 1.3.2, we obtain

$$\Sigma_{n-1}\cup\{\sigma_2,\dots,\sigma_r\}\vdash(\sigma_1\to(\alpha\wedge\neg\alpha)),$$

and, by use of the tautology

$$((\beta\to\gamma)\to(\alpha\wedge\neg\alpha))\to(\beta\wedge\neg\gamma)$$

and (MP) it follows, finally, that

$$\Sigma_{n-1}\cup\{\sigma_2,\dots,\sigma_r\}\vdash(\exists x_1\,\varphi_1\wedge\neg\varphi_1(x_1/c_{k_1})).$$

If one bears in mind that $\exists x_1$ is an abbreviation for $\neg\forall x_1\neg$, then we obtain via ($\wedge$ B$_1$), on the one hand,

$$\Sigma_{n-1}\cup\{\sigma_2,\dots,\sigma_r\}\vdash\neg\forall x_1\neg\varphi_1,\tag{1.4.3.7}$$

and, via ($\wedge$B$_2$) on the other hand,

$$\Sigma_{n-1}\cup\{\sigma_2,\dots,\sigma_r\}\vdash\neg\varphi_1(x_1/c_{k_1}).\tag{1.4.3.8}$$

The provabilities asserted in (1.4.3.7) and (1.4.3.8) are meant in the language $L^{(2)}$. If we now define $L^{(1)}$ by

$$I^{(1)}=I_n,\quad J^{(1)}=J_n,\quad \text{and}\quad K^{(1)}=K_{n-1}\cup\{g_n^{-1}(\exists x_2\,\varphi_2),\dots,g_n^{-1}(\exists x_r\,\varphi_r)\},$$

then we recognize that $\neg\forall x_1\neg\varphi_1$ as well as the set of sentences

$$\Pi:=\Sigma_{n-1}\cup\{\sigma_2,\dots,\sigma_r\}$$

already lie in Sent$(L^{(1)})$. Applying (a suitable version of) Lemma 1.4.3 to the proofs whose existence is asserted by (1.4.3.7) and (1.4.3.8), we obtain, on the one hand, a deduction

$$\Pi\vdash\neg\forall x_1\neg\varphi_1$$

in $L^{(1)}$, and, on the other hand, a deduction

$$\Pi \vdash \neg\varphi_1(x_1/c_{k_1})(c_{k_1}/y)$$

likewise in $L^{(1)}$. Here $y$ is a suitably chosen "new" variable (i.e. a variable not occurring in the proof of $\neg\varphi_1(x_1/c_{k_1})$ or in the proof of $\neg\forall x_1 \neg\varphi_1$ from $\Pi$, to which we applied Lemma 1.4.3). Now it is obvious that

$$\varphi_1(x_1/c_{k_1})(c_{k_1}/y) \text{ is } \varphi_1(x_1/y),$$

since $\varphi_1$ is a member of $\mathrm{Fml}(L_{n-1})$. We therefore have

$$\Pi \vdash \neg\varphi_1(x_1/y).$$

Via an application of ($\forall$) on $y$ and ($\forall$B) we obtain, first,

$$\Pi \vdash \forall y \neg\varphi_1(x_1/y)$$

and then

$$\Pi \vdash \neg\varphi_1(x_1/y)(y/x_1).$$

If one considers that $y$ is new for $\varphi_1$, then one understands immediately that

$$\varphi_1(x_1/y)(y/x_1) \text{ is } \varphi_1.$$

Thus we have $\Pi \vdash \neg\varphi_1$ and thus we finally obtain

$$\Pi \vdash \forall x_1 \neg\varphi_1.$$

This derivability, together with the derivability

$$\Pi \vdash \neg\forall x_1 \neg\varphi_1$$

shows that $\Pi$ is inconsistent in $L^{(1)}$.

Just as we have reduced the inconsistency of $\Sigma_{n-1} \cup \{\sigma_1, \ldots, \sigma_r\}$ to that of $\Sigma_{n-1} \cup \{\sigma_2, \ldots, \sigma_r\}$, we can, through iteration, finally deduce a contradiction already in $\Sigma_{n-1}$. Since this contradicts our hypothesis, the consistency of $\Sigma_n$ follows. In this way, all the $\Sigma_n$ are recognized as consistent.

Now the consistency of $\Sigma' = \bigcup_{n \in \mathbb{N}} \Sigma_n$ is seen as follows: since the proof of a contradiction from $\Sigma'$ is a finite sequence of formulae, and both the languages $L_n$ as well as the sets $\Sigma_n$ form ascending chains, there is an $n \in \mathbb{N}$ such that this proof is already a proof from $\Sigma_n$ in the language $L_n$. This is, however, impossible, because of the previously proved consistency of $\Sigma_n$. □

We shall carry out the *second step* in determining a domain in which all sentences in our consistent set $\Sigma$ will hold (1.4.1.3), in the extension language $L'$ constructed just above. For this we shall apply the following theorem, which we formulate for an arbitrary language (again denoted by $L$).

**Theorem 1.4.4.** *To each consistent set* $\Sigma \subseteq \mathrm{Sent}(L)$ *there is a maximal consistent extension* $\Sigma^* \subseteq \mathrm{Sent}(L)$ *of* $\Sigma$*; this means that* $\Sigma \subseteq \Sigma^*$*,* $\Sigma^*$ *is consistent, and whenever* $\Sigma^* \subseteq \Sigma_1 \subseteq \mathrm{Sent}(L)$ *and* $\Sigma_1$ *is consistent,* $\Sigma^* = \Sigma_1$.

*Proof*: We consider the system

$$\mathfrak{M} = \{\Sigma_1 \subseteq \mathrm{Sent}(L) \mid \Sigma \subseteq \Sigma_1,\ \Sigma_1 \text{ consistent}\}.$$

Since $\Sigma \in \mathfrak{M}$, $\mathfrak{M}$ is not empty. If we are given a subsystem $\mathfrak{M}' \subseteq \mathfrak{M}$ that is linearly ordered by inclusion (i.e. for $\Sigma_1, \Sigma_2 \in \mathfrak{M}'$, either $\Sigma_1 \subseteq \Sigma_2$ or $\Sigma_2 \subseteq \Sigma_1$), then the set

$$\Sigma' := \bigcup_{\Sigma_1 \in \mathfrak{M}'} \Sigma_1$$

is clearly an upper bound for $\mathfrak{M}'$ in $\mathfrak{M}$. To see this, note first that for each $\Sigma_1 \in \mathfrak{M}'$, $\Sigma_1 \subseteq \Sigma'$, trivially. The consistency of $\Sigma'$ rests simply on the finiteness of a proof of a hypothetical contradiction from $\Sigma'$: such a proof can utilize only finitely many axioms $\sigma_1, \ldots, \sigma_n \in \Sigma'$. Each of the $\sigma_i$ lies in a member of the system $\mathfrak{M}'$, say, $\sigma_i \in \Sigma_i$. Since, however, the sets $\Sigma_1, \ldots, \Sigma_n$ are comparable, one of them, say $\Sigma_n$, must contain all the others as subsets. Then the hypothetical contradiction would be deducible already from $\Sigma_n$, which is impossible.

We have therefore shown that the system $\mathfrak{M}$ fulfils the hypotheses of Zorn's lemma. Therefore there is a maximal element $\Sigma^*$ in $\mathfrak{M}$. Then, according to the definition of $\mathfrak{M}$, $\Sigma \subseteq \Sigma^*$ and $\Sigma^*$ is consistent. □

**Remark 1.4.5** *If the language L is countable, i.e. the sets I, J and K are (finite or) countable, then Zorn's lemma can be avoided in the proof of the above lemma.*

*Proof* (of 1.4.5): In this case we can start with an enumeration $(\varphi_n)_{n \in \mathbb{N}}$ of all $L$-sentences. Then we recursively define

$$\Sigma_0 = \Sigma$$

$$\Sigma_{n+1} = \begin{cases} \Sigma_n & \text{in case } \Sigma_n \cup \{\varphi_n\} \text{ is inconsistent, and} \\ \Sigma_n \cup \{\varphi_n\} & \text{otherwise.} \end{cases} \tag{1.4.5.1}$$

Thus we obtain an ascending chain

$$\Sigma_0 \subseteq \Sigma_1 \subseteq \cdots \subseteq \Sigma_n \subseteq \Sigma_{n+1} \subseteq \cdots$$

of consistent sets of sentences. From this it follows, as before, that also

$$\Sigma^* := \bigcup_{n \in \mathbb{N}} \Sigma_n$$

is consistent. Because of $\Sigma = \Sigma_0 \subseteq \Sigma^*$, it remains only to show the maximality of $\Sigma^*$. Assume there were a $\varphi \in \text{Sent}(L)$ such that $\Sigma^* \cup \{\varphi\}$ were still consistent. This sentence $\varphi$ occurs in the enumeration $(\varphi_n)_{n\in\mathbb{N}}$ of all $L$-sentences. Let us say $\varphi$ is $\varphi_n$. From the consistency of $\Sigma^* \cup \{\varphi\}$ follows that of $\Sigma^* \cup \{\varphi_n\}$ and, *a fortiori*, that of $\Sigma_n \cup \{\varphi_n\}$. Therefore

$$\Sigma_n \cup \{\varphi_n\} = \Sigma_{n+1} \subseteq \Sigma^*,$$

by (1.4.5.1). From this follows $\varphi_n \in \Sigma^*$. Thus $\Sigma^*$ is maximal (as well as consistent). □

Now we apply Theorem 1.4.4 to the consistent set $\Sigma' \subseteq \text{Sent}(L')$ obtained in Theorem 1.4.2, in order to obtain a maximal consistent extension $\Sigma^* \subseteq \text{Sent}(L')$ of $\Sigma'$. For such a $\Sigma^*$ we have:

(I) $\Sigma^*$ is maximal consistent in $\text{Sent}(L')$;
(II) for each $\exists x\,\varphi$ in $\text{Sent}(L')$ there is a $k \in K'$ with $(\exists x\,\varphi \to \varphi(x/c_k))$ in $\Sigma^*$.

These two properties of $\Sigma^*$ canonically determine a domain in which all sentences $\sigma \in \Sigma^*$ hold, as we shall see. In particular, all $\sigma \in \Sigma$ will hold there.

We first consider the set of *constant $L'$-terms*:

$$\text{CT} := \{t \in \text{Tm}(L') \mid \text{no variable occurs in } t\}. \tag{1.4.5.2}$$

CT contains, in particular, all $c_k$ with $k \in K'$. We define a binary relation $\approx$ on CT: for $t_1, t_2 \in \text{CT}$, we set

$$t_1 \approx t_2 \quad \text{iff} \quad \Sigma^* \vdash t_1 \doteq t_2. \tag{1.4.5.3}$$

With the help of axiom (I1) (1.3.0.9) and Rules (S) and (Tr), we recognize immediately that $\approx$ is an equivalence relation on CT, i.e. for $t_1, t_2, t_3 \in \text{CT}$:

(i) $t_1 \approx t_1$,
(ii) if $t_1 \approx t_2$, then $t_2 \approx t_1$, and
(iii) if $t_1 \approx t_2$ and $t_2 \approx t_3$, then $t_1 \approx t_3$.

Now the sought-for domain is the set

$$A := \text{CT}/\approx \tag{1.4.5.4}$$

of all equivalence classes $\bar{t}$ of constant terms. Here, as usual, we define for $t \in \text{CT}$:

$$\bar{t} = \{t_1 \in \text{CT} \mid t \approx t_1\}. \tag{1.4.5.5}$$

Then

$$\overline{t_1} = \overline{t_2} \quad \text{iff} \quad t_1 \approx t_2. \tag{1.4.5.6}$$

In order to be able to speak meaningfully of the truth of a sentence in the domain (something that we shall make precise only in the next section), we must say which relations, functions and individuals the symbols $R_i$, $f_j$ and $c_k$ name – i.e. we must give an interpretation of these symbols.

To each $i \in I$ we define a $\lambda(i)$-place *relation* $\mathcal{R}_i$ on the domain $A$, by declaring, for term-classes $\overline{t_1}, \ldots, \overline{t_{\lambda(i)}}$, that

$$\mathscr{R}_i(\overline{t_1},\ldots,\overline{t_{\lambda(i)}}) \text{ holds if and only if } \Sigma^* \vdash R_i(t_1,\ldots,t_{\lambda(i)}). \tag{1.4.5.7}$$

Here the notation $\mathscr{R}_i(\overline{t_1},\ldots,\overline{t_{\lambda(i)}})$ means, as usual, that the relation $\mathscr{R}_i$ holds at the $\lambda(i)$-tuple $(\overline{t_1},\ldots,\overline{t_{\lambda(i)}})$ of term-classes; i.e. that $(\overline{t_1},\ldots,\overline{t_{\lambda(i)}}) \in \mathscr{R}_i$. One should observe, however, that the above definition of $\mathscr{R}_i$ refers back to a particular choice of *representatives* $t_\nu$ of the term-classes $\overline{t_\nu}$. It must be shown that a choice of other representatives leads to the same definition. Thus, suppose $t_1 \approx t_1'$, ..., $t_{\lambda(i)} \approx t'_{\lambda(i)}$; then we must show:

$$\Sigma^* \vdash R_i(t_1,\ldots,t_{\lambda(i)}) \quad \text{iff} \quad \Sigma^* \vdash R_i(t_1',\ldots,t'_{\lambda(i)}). \tag{1.4.5.8}$$

By symmetry, it obviously suffices to show only one direction. So let us assume that

$$\Sigma^* \vdash R_i(t_1,\ldots,t_{\lambda(i)}).$$

Along with this deducibility we have, according to the hypothesis, the deducibilities

$$\Sigma^* \vdash t_\nu \doteq t_\nu', \quad \text{for } 1 \le \nu \le \lambda(i).$$

By piecing these $\lambda(i)+1$ deductions together, we assemble a deduction from $\Sigma^*$ that ends with the following lines:

$$\begin{gathered} t_1 \doteq t_1' \\ \vdots \\ t_{\lambda(i)} \doteq t'_{\lambda(i)} \\ R_i(t_1,\ldots,t_{\lambda(i)}). \end{gathered}$$

Now we extend this proof with the following lines:

$$\begin{aligned} R_i(t_1,t_2,\ldots,t_{\lambda(i)}) &\to R_i(t_1',t_2,\ldots,t_{\lambda(i)}) \\ &\phantom{\to} R_i(t_1',t_2,\ldots,t_{\lambda(i)}) \\ R_i(t_1',t_2,t_3,\ldots,t_{\lambda(i)}) &\to R_i(t_1',t_2',t_3,\ldots,t_{\lambda(i)}) \\ &\phantom{\to} R_i(t_1',t_2',t_3,\ldots,t_{\lambda(i)}) \\ \vdots \qquad & \vdots \qquad \vdots \\ R_i(t_1',t_2',\ldots,t'_{\lambda(i)-1},t_{\lambda(i)}) &\to R_i(t_1',t_2',\ldots,t'_{\lambda(i)-1},t'_{\lambda(i)}) \\ &\phantom{\to} R_i(t_1',t_2',\ldots,t'_{\lambda(i)-1},t'_{\lambda(i)}). \end{aligned}$$

These lines arise from alternating application of ($R_i$) (p. 20) and (MP). Altogether we obtain

$$\Sigma^* \vdash R_i(t_1',\ldots,t'_{\lambda(i)}),$$

proving (1.4.5.8).

Next, for each $j \in J$ we define a $\mu(j)$-place *function* $\mathscr{F}_j : A^{\mu(j)} \to A$ by defining, for term-classes $\overline{t_1}, \ldots, \overline{t_{\mu(j)}}$,

$$\mathscr{F}_j(\overline{t_1}, \ldots, \overline{t_{\mu(j)}}) := \overline{f_j(t_1, \ldots, t_{\mu(j)})}. \tag{1.4.5.9}$$

Here, too, we must show that this definition does not depend on the choice of representative $t_\nu$ of the class $\overline{t_\nu}$. By application of the Rules ($\mathrm{F_j}$) and (Tr) (p. 20), we obtain, following the above pattern of argument, a proof from $\Sigma^*$ of

$$f_j(t_1, \ldots, t_{\mu(j)}) \doteq f_j(t'_1, \ldots, t'_{\mu(j)}),$$

if one assumes

$$\Sigma^* \vdash t_\nu \doteq t'_\nu \quad \text{for } 1 \leq \nu \leq \mu(j).$$

Finally, for every $k \in K$,

$$\text{we take the class } \overline{c_k} \text{ to be the interpretation of } c_k. \tag{1.4.5.10}$$

The content of the next theorem is that, under the above interpretations, all $L'$-sentences $\sigma \in \Sigma^*$, and only those, hold in the domain $A$. This will become conclusively clear, however, only after we have, in the next section, made the notion of satisfaction precise. We place a small technical lemma before the promised theorem.

**Lemma 1.4.6.** *The maximal consistent set $\Sigma^*$ of sentences is deductively closed; i.e. for each $\alpha \in \mathrm{Sent}(L')$ with $\Sigma^* \vdash \alpha$, $\alpha \in \Sigma^*$.*

*Proof*: In view of the maximal consistency of $\Sigma^*$, it suffices to show that $\Sigma^* \cup \{\alpha\}$ is consistent whenever $\Sigma^* \vdash \alpha$. But this is clear: namely, if $\Sigma^* \cup \{\alpha\}$ were inconsistent, then we would have, in particular,

$$\Sigma^* \cup \{\alpha\} \vdash \neg\alpha,$$

which, together with the Deduction Theorem 1.3.2, would lead to

$$\Sigma^* \vdash (\alpha \to \neg\alpha).$$

Because of the tautology

$$(\alpha \to \neg\alpha) \to \neg\alpha,$$

this would lead, finally, to $\Sigma^* \vdash \neg\alpha$, contrary to the assumption that $\Sigma^*$ is consistent. □

**Theorem 1.4.7.** *Suppose $\Sigma^*$ is an arbitrary subset of* $\mathrm{Sent}(L')$ *possessing properties* (I) *and* (II) *(p. 33). Then, for every $\alpha$, $\beta$, and $\forall x\,\varphi$ in* $\mathrm{Sent}(L')$, *we have:*

(a) $\neg\alpha \in \Sigma^*$ *iff* $\alpha \notin \Sigma^*$;

(b) $(\alpha \wedge \beta) \in \Sigma^*$ *iff* ($\alpha \in \Sigma^*$ *and* $\beta \in \Sigma^*$); *and*

(c) $\forall x\,\varphi \in \Sigma^*$ *iff* $\varphi(x/t) \in \Sigma^*$ *for all* $t \in \mathrm{CT}$ (1.4.5.2).

*Proof*: (a) ($\Rightarrow$) Since $\Sigma^*$ is consistent (I), $\alpha$ and $\neg\alpha$ cannot both lie in $\Sigma^*$.

($\Leftarrow$) From $\alpha \notin \Sigma^*$ we deduce immediately that $\Sigma^* \nvdash \alpha$, by Lemma 1.4.6. From this it follows that $\Sigma^* \cup \{\neg\alpha\}$ is consistent, by Lemma 1.4.1. But since $\Sigma^*$ is maximal consistent (I), $\Sigma^* \cup \{\neg\alpha\} = \Sigma^*$, whence $\neg\alpha \in \Sigma^*$.

(b) ($\Rightarrow$) From $(\alpha \wedge \beta) \in \Sigma^*$ it follows trivially that $\Sigma^* \vdash (\alpha \wedge \beta)$. Then $\Sigma^* \vdash \alpha$ and $\Sigma^* \vdash \beta$, by Rules ($\wedge$ B$_1$) and ($\wedge$ B$_2$). Now use Lemma 1.4.6.

($\Leftarrow$) Use Rule ($\wedge$) (p. 19) followed by Lemma 1.4.6.

(c) ($\Rightarrow$) If $\forall x \varphi \in \Sigma^*$, then $\Sigma^* \vdash \forall x \varphi$. Then for any $t \in \mathrm{CT}$, $\Sigma^* \vdash \varphi(x/t)$, by Rule ($\forall$ B) (p. 18), which applies here since every constant term $t$ is, vacuously, free for $x$ in $\varphi$. By (1.4.6), we get $\varphi(x/t) \in \Sigma^*$.

($\Leftarrow$) Assume $\forall x \varphi \notin \Sigma^*$. Then $\neg\forall x \varphi \in \Sigma^*$, by (a). From this we would like to deduce $\exists x \neg\varphi \in \Sigma^*$. $\Sigma^* \vdash (\neg\neg\varphi \rightarrow \varphi)$, since $\neg\neg\varphi \rightarrow \varphi$ is a tautology. From this, $\Sigma^* \cup \{\neg\neg\varphi\} \vdash \varphi$ follows immediately, and thence (with Lemma 1.3.1)

$$\Sigma^* \cup \{\forall x \neg\neg\varphi\} \vdash \forall x \varphi.$$

Since $\forall x \varphi$ is a sentence by hypothesis, so is $\forall x \neg\neg\varphi$. Therefore we may apply the Deduction Theorem 1.3.2 to obtain

$$\Sigma^* \vdash (\forall x \neg\neg\varphi \rightarrow \forall x \varphi).$$

From this we obtain, using (CP) (p. 18),

$$\Sigma^* \vdash (\neg\forall x \varphi \rightarrow \neg\forall x \neg\neg\varphi),$$

and, using $\neg\forall x \varphi \in \Sigma^*$, finally $\exists x \neg\varphi \in \Sigma^*$. By hypothesis (II) there is at least one $t \in \mathrm{CT}$ (in fact, $t$ may even be taken to be a constant *symbol*) such that

$$\Sigma^* \vdash (\exists x \neg\varphi \rightarrow \neg\varphi(x/t)).$$

Therefore for this $t$ we conclude $\neg\varphi(x/t) \in \Sigma^*$. But then $\varphi(x/t) \notin \Sigma^*$, by (I). □

## 1.5 First-Order Semantics

In this section we want to define what it means for a formula $\varphi$ of a formal language $L$ to hold in, or to be satisfied by, a particular mathematical structure, and, more generally, what it should mean for such a structure to be a model of an axiom system $\Sigma$. In order to be able to carry this out meaningfully, we must first fix the boundaries of a domain of objects to which our quantifiers $\forall u$ and $\exists v$ should refer, i.e. over which the variables $v_0, v_1, \ldots$ should "vary". After that we must define the interpretation of each relation symbol $R_i$ ($i \in I$), each function symbol $f_j$ ($j \in J$), and each constant symbol $c_k$ ($k \in K$).

Suppose we are given a formal language $L = (\lambda, \mu, K)$. An *L-structure* $\mathfrak{A}$ is determined by the following data:

$|\mathfrak{A}|$: a nonempty set, the *universe* of $\mathfrak{A}$;

$R_i^{\mathfrak{A}}$: a $\lambda(i)$-place relation on $|\mathfrak{A}|$ (i.e. a subset of $|\mathfrak{A}|^{\lambda(i)}$), for each $i \in I$;

$f_j^{\mathfrak{A}}$: a $\mu(j)$-place function defined on all of $|\mathfrak{A}|$ (i.e. a function $|\mathfrak{A}|^{\mu(j)} \to |\mathfrak{A}|$), for each $j \in J$;

$c_k^{\mathfrak{A}}$: a fixed element of $|\mathfrak{A}|$, for each $k \in K$.

We summarize this with the notation

$$\mathfrak{A} = \langle |\mathfrak{A}|;\ (R_i^{\mathfrak{A}})_{i\in I};\ (f_j^{\mathfrak{A}})_{j\in J};\ (c_k^{\mathfrak{A}})_{k\in K} \rangle.$$

If we again consider the language $L$ (p. 21) used in our example of a formal proof in Section 1.3, with the relation symbol $\leq$, the function symbols $-$, $+$, $\cdot$, and the constant symbol 0, then the following is an $L$-structure:

$$\mathfrak{R} = \langle \mathbb{R};\ \leq^{\mathbb{R}};\ -^{\mathbb{R}},\ +^{\mathbb{R}},\ \cdot^{\mathbb{R}};\ 0^{\mathbb{R}} \rangle. \tag{1.5.0.1}$$

Here $\mathbb{R}$ is the set of real numbers; $\leq^{\mathbb{R}}$ is the usual "less-than-or-equal-to" relation on $\mathbb{R}$; $-^{\mathbb{R}}$, $+^{\mathbb{R}}$, and $\cdot^{\mathbb{R}}$ are the usual operations "negative" (unary), "plus" (binary), and "times" (binary) on $\mathbb{R}$; and $0^{\mathbb{R}}$ is the real number "zero".

Let us pursue this example further by considering the formula

$$\exists v_0\,(0 \leq v_0 \wedge v_0 \leq v_1). \tag{1.5.0.2}$$

Then the question whether this formula holds in (or is satisfied by) $\mathfrak{R}$ – presupposing for now a definition of satisfaction that agrees with our intuition – can be meaningfully answered only after we assign to $v_1$ a definite value in $\mathbb{R}$: for a negative value of $v_1$, the answer is no; for other values, the answer is yes. Thus we see in this example that for a meaningful definition of the satisfaction of a formula $\varphi$ in $\mathfrak{R}$, each of the free variables of $\varphi$ must be assigned a value in $\mathbb{R}$. For certain technical reasons we assign values not only to the free variables of one formula, but to the free variables of all formulae; i.e. to all variables. However, in the definition of satisfaction we must make sure that in the case of the bound variables of the formula $\varphi$ under consideration, the fixed assignment by $h$ is "unfixed".

An assignment of values in $|\mathfrak{A}|$ for all variables will be called an *evaluation of the variables in* $\mathfrak{A}$. Thus, an evaluation in $\mathfrak{A}$ is a function

$$h : \mathrm{Vbl} \to |\mathfrak{A}|.$$

If $h$ is an evaluation in $\mathfrak{A}$, then the value $h(x) \in |\mathfrak{A}|$ is assigned to the variable $x$. For each $a \in |\mathfrak{A}|$, each $x \in \mathrm{Vbl}$, and each evaluation $h$, the function $h\binom{x}{a}$, defined as follows, is again an evaluation:

$$h\binom{x}{a}(v) = \begin{cases} h(v) & \text{for } v \neq x \\ a & \text{for } v = x. \end{cases} \tag{1.5.0.3}$$

The evaluations $h$ and $h\binom{x}{a}$ agree with each other at all variables other than $x$. At $x$, the value of $h$ is $h(x)$, while that of $h\binom{x}{a}$ is $a$. Obviously, $h\binom{x}{h(x)} = h$.

Next we define, by recursion on the construction of a term, the *value* $t^{\mathfrak{A}}[h]$ *of the term* $t$ *under the evaluation* $h$ (or simply the *h-value of* $t$) in $\mathfrak{A}$:

$$\begin{aligned} v^{\mathfrak{A}}[h] &:= h(v) \\ c_k^{\mathfrak{A}}[h] &:= c_k^{\mathfrak{A}} \\ f_j(t_1,\ldots,t_{\mu(j)})^{\mathfrak{A}}[h] &:= f_j^{\mathfrak{A}}(t_1^{\mathfrak{A}}[h],\ldots,t_{\mu(j)}^{\mathfrak{A}}[h]). \end{aligned} \tag{1.5.0.4}$$

It is clear that these equations determine the $h$-value (in $|\mathfrak{A}|$) of each term, by starting the recursive process with the $h$-values of its simplest subterms – the variables and constant symbols occurring in it.

The *satisfaction of a formula* $\varphi$ *under an evaluation* $h$ *in* $\mathfrak{A}$ will be a ternary relation of our metatheory. If this relation between $\mathfrak{A}$, $\varphi$, and $h$ holds, then we shall write $\mathfrak{A} \models \varphi[h]$ (pronounced: "$\varphi$ holds in $\mathfrak{A}$ under $h$", "$\varphi$ is true in $\mathfrak{A}$ under $h$", or "$\varphi$ is satisfied by $\mathfrak{A}$ under $h$"); if this relation does not hold, then we shall write $\mathfrak{A} \not\models \varphi[h]$. This relation will likewise be defined by recursion on the construction of formulae, starting with the simplest formulae, the atomic formulae, and indeed simultaneously for all evaluations.

For atomic formulae $t_1 \doteq t_2$ and $R_i(t_1,\ldots,t_{\lambda(i)})$, we declare, for an arbitrary evaluation $h$ in $\mathfrak{A}$:

$$\mathfrak{A} \models t_1 \doteq t_2\,[h] \quad \text{iff} \quad t_1^{\mathfrak{A}}[h] = t_2^{\mathfrak{A}}[h]; \tag{1.5.0.5}$$

$$\mathfrak{A} \models R_i(t_1,\ldots,t_{\lambda(i)})\,[h] \text{ iff } R_i^{\mathfrak{A}}(t_1^{\mathfrak{A}}[h],\ldots,t_{\lambda(i)}^{\mathfrak{A}}[h]). \tag{1.5.0.6}$$

Thereafter, for formulae $\varphi$ and $\psi$ we continue our recursive definition as follows:

$$\mathfrak{A} \models \neg\varphi\,[h] \quad \text{iff} \quad \mathfrak{A} \not\models \varphi[h]; \tag{1.5.0.7}$$

$$\mathfrak{A} \models (\varphi \wedge \psi)\,[h] \quad \text{iff} \quad (\mathfrak{A} \models \varphi[h] \text{ and } \mathfrak{A} \models \psi[h]); \tag{1.5.0.8}$$

$$\mathfrak{A} \models \forall x\varphi\,[h] \quad \text{iff} \quad \mathfrak{A} \models \varphi\left[h\binom{x}{a}\right], \text{ for all } a \in |\mathfrak{A}|. \tag{1.5.0.9}$$

Observe that in the last case, where $x$ is certainly bound, the prescription by $h$ of a particular value in $|\mathfrak{A}|$ for $x$ is unfixed, since in this case we consider, instead of $h$ itself, an alteration of $h$ at the point $x$; here every such alteration is taken into consideration. In this way we ensure that the definition of satisfaction really agrees with our intuition.

Using the definitions of $\vee$, $\rightarrow$, $\leftrightarrow$, and $\exists$ from Section 1.2, and the definition of satisfaction, one obtains immediately the following equivalences:

$$\begin{array}{lll} \mathfrak{A} \models (\varphi \vee \psi)\,[h] & \text{iff} \quad (\mathfrak{A} \models \varphi\,[h] \quad \text{or} \quad \mathfrak{A} \models \psi\,[h]); & \\ \mathfrak{A} \models (\varphi \rightarrow \psi)\,[h] & \text{iff} \quad (\mathfrak{A} \models \varphi\,[h] \text{ implies } \mathfrak{A} \models \psi\,[h]); & (1.5.0.10) \\ \mathfrak{A} \models (\varphi \leftrightarrow \psi)\,[h] & \text{iff} \quad (\mathfrak{A} \models \varphi\,[h] \quad \text{iff} \quad \mathfrak{A} \models \psi\,[h]); & \\ \mathfrak{A} \models \exists x \varphi\,[h] & \text{iff there is an } a \in |\mathfrak{A}| \text{ such that } \mathfrak{A} \models \varphi\left[h\binom{x}{a}\right]. & (1.5.0.11) \end{array}$$

Here, the words "and", "or", "implies", "iff", "for all $a$" and "there exists an $a$" are to be understood in the usual mathematical sense; in particular, "or" is not used in the exclusive sense, and "implies" is regarded as false only when the premise is true and the conclusion is false; cf. the truth table on page 16.

For the structure $\mathfrak{R}$ in (1.5.0.1) above (with the set of real numbers as universe), and the formula $\exists v_0\,(0 \leq v_0 \,\wedge\, v_0 \leq v_1)$ (1.5.0.2), we have the following translation:

$$\begin{aligned} &\mathfrak{R} \models \exists v_0\,(0 \leq v_0 \,\wedge\, v_0 \leq v_1)\,[h] \\ &\qquad \text{iff there exists an } a \in \mathbb{R} \text{ such that } \mathfrak{R} \models (0 \leq v_0 \,\wedge\, v_0 \leq v_1)\left[h\binom{v_0}{a}\right] \\ &\qquad \text{iff there exists an } a \in \mathbb{R} \text{ such that } \left(0 \leq^{\mathfrak{R}} a \,\wedge\, a \leq^{\mathfrak{R}} h(v_1)\right). \end{aligned}$$

This example shows yet again that the definition of satisfaction has fulfilled its purpose: it translates, using the prescription of values of the free variables given by an evaluation $h$, a string $\varphi$ of symbols into an assertion in the metalanguage – the interpretation of $\varphi$ in $\mathfrak{R}$ under $h$.

In this example we see, in addition, that for the relation $\models$ to hold, the only variables whose $h$-values are material are the free variables of $\varphi$. This can be shown in general:

**Lemma 1.5.1.** *Let $h'$ and $h''$ be evaluations in the L-structure $\mathfrak{A}$. Then:*
*(a) If $h'$ and $h''$ agree on the variables in the L-term $t$, then $t^{\mathfrak{A}}[h'] = t^{\mathfrak{A}}[h'']$.*
*(b) If $h'$ and $h''$ agree on the free variables occurring in the L-formula $\varphi$, then*

$$\mathfrak{A} \models \varphi\,[h'] \quad \textit{iff} \quad \mathfrak{A} \models \varphi\,[h'']. \tag{1.5.1.1}$$

*Proof*: (a) The following equations prove this by induction on the recursive construction of terms:

$$\begin{aligned} v^{\mathfrak{A}}[h'] &= h'(v) = h''(v) = v^{\mathfrak{A}}[h''] \\ c_k^{\mathfrak{A}}[h'] &= c_k^{\mathfrak{A}} = c_k^{\mathfrak{A}}[h''] \\ f_j(t_1,\ldots,t_{\mu(j)})^{\mathfrak{A}}[h'] &= f_j^{\mathfrak{A}}(t_1^{\mathfrak{A}}[h'],\ldots,t_{\mu(j)}^{\mathfrak{A}}[h']) \\ &= f_j^{\mathfrak{A}}(t_1^{\mathfrak{A}}[h''],\ldots,t_{\mu(j)}^{\mathfrak{A}}[h'']) \\ &= f_j(t_1,\ldots,t_{\mu(j)})^{\mathfrak{A}}[h'']. \end{aligned}$$

(b) We prove (1.5.1.1) similarly, using (a) and induction on the recursive construction of formulae:

$$\begin{aligned}
\mathfrak{A} \models t_1 \doteq t_2\,[h'] \quad &\text{iff} \quad t_1^{\mathfrak{A}}\,[h'] = t_2^{\mathfrak{A}}\,[h'] \\
&\text{iff} \quad t_1^{\mathfrak{A}}\,[h''] = t_2^{\mathfrak{A}}\,[h''] \\
&\text{iff} \quad \mathfrak{A} \models t_1 \doteq t_2\,[h''].
\end{aligned}$$

$$\begin{aligned}
\mathfrak{A} \models R_i(t_1,\ldots)\,[h'] \ &\text{iff} \quad R_i^{\mathfrak{A}}(t_1^{\mathfrak{A}}\,[h'],\ldots) \\
&\text{iff} \quad R_i^{\mathfrak{A}}(t_1^{\mathfrak{A}}\,[h''],\ldots) \\
&\text{iff} \quad \mathfrak{A} \models R_i(t_1,\ldots)\,[h''].
\end{aligned}$$

$$\begin{aligned}
\mathfrak{A} \models \neg\varphi\,[h'] \quad &\text{iff} \quad \mathfrak{A} \not\models \varphi\,[h'] \\
&\text{iff} \quad \mathfrak{A} \not\models \varphi\,[h''] \quad \text{(ind. hyp.)} \\
&\text{iff} \quad \mathfrak{A} \models \neg\varphi\,[h''].
\end{aligned}$$

$$\begin{aligned}
\mathfrak{A} \models (\varphi \wedge \psi)\,[h'] \quad &\text{iff} \quad (\mathfrak{A} \models \varphi\,[h'] \text{ and } \mathfrak{A} \models \psi\,[h']) \\
&\text{iff} \quad (\mathfrak{A} \models \varphi\,[h''] \text{ and } \mathfrak{A} \models \psi\,[h'']) \quad \text{(ind. hyp.)} \\
&\text{iff} \quad \mathfrak{A} \models (\varphi \wedge \psi)\,[h''].
\end{aligned}$$

$$\begin{aligned}
\mathfrak{A} \models \forall x\varphi\,[h'] \quad &\text{iff} \quad \mathfrak{A} \models \varphi\left[h'\tbinom{x}{a}\right] \quad \text{for all } a \in |\mathfrak{A}| \\
&\text{iff} \quad \mathfrak{A} \models \varphi\left[h''\tbinom{x}{a}\right] \quad \text{for all } a \in |\mathfrak{A}| \qquad (1.5.1.2) \\
&\text{iff} \quad \mathfrak{A} \models \forall x\varphi\,[h''].
\end{aligned}$$

In (1.5.1.2) we applied the inductive hypothesis to the shorter formula $\varphi$ and the evaluations $h'\binom{x}{a}$ and $h''\binom{x}{a}$. Note that these two evaluations agree with each other on all free variables of $\varphi$, due to the common alteration of $h'$ and $h''$ at $x$. □

From Lemma 1.5.1 we see, in particular, that the satisfaction of a sentence $\varphi$ in $\mathfrak{A}$ does not depend on the evaluation $h$ considered. That is, for $\varphi \in \mathrm{Sent}(L)$ and evaluations $h'$ and $h''$ in $\mathfrak{A}$, we always have

$$\mathfrak{A} \models \varphi\,[h'] \quad \text{iff} \quad \mathfrak{A} \models \varphi\,[h'']. \qquad (1.5.1.3)$$

We say that a formula $\varphi$ *holds in*, or *is true in*, or *is satisfied by*, $\mathfrak{A}$ (and we write $\mathfrak{A} \models \varphi$), in case $\mathfrak{A} \models \varphi\,[h]$ holds for all evaluations $h$ in $\mathfrak{A}$. One easily sees that

$$\mathfrak{A} \models \varphi \quad \text{iff} \quad \mathfrak{A} \models \forall\varphi \ \text{(p. 13)}.$$

This follows, by induction, from the equivalence

$$\mathfrak{A} \models \varphi \quad \text{iff} \quad \mathfrak{A} \models \forall x\varphi \qquad (1.5.1.4)$$

(for any variable $x$), which is proved as follows:

$$\begin{aligned}\mathfrak{A} \models \varphi \quad &\text{iff} \quad \mathfrak{A} \models \varphi\,[h] \quad \text{for every evaluation } h\\ &\text{iff} \quad \mathfrak{A} \models \varphi\left[h\tbinom{x}{a}\right] \text{ for every evaluation } h \text{ and every } a \in |\mathfrak{A}|\\ &\text{iff} \quad \mathfrak{A} \models \forall x\,\varphi\,[h] \quad \text{for every evaluation } h\\ &\text{iff} \quad \mathfrak{A} \models \forall x\,\varphi.\end{aligned}$$

If $\Sigma$ is a set of $L$-sentences, then an $L$-structure $\mathfrak{A}$ is called a *model* of $\Sigma$ if every $\sigma \in \Sigma$ holds in $\mathfrak{A}$, i.e. $\mathfrak{A} \models \sigma$ for all $\sigma \in \Sigma$. In this case we write $\mathfrak{A} \models \Sigma$.

Now we come back to the starting point (1.4.0.1) of our inquiry in Section 1.4: How can an "undeducibility" $\Sigma \nvdash \varphi$ occur? The answer is given by the following theorem (which is the promised, precise version of (1.4.0.2)):

**Theorem 1.5.2** (Gödel's completeness theorem). *Let* $\Sigma \subseteq \mathrm{Sent}(L)$ *and* $\varphi \in \mathrm{Sent}(L)$. *Then from* $\Sigma \nvdash \varphi$ *follows the existence of a "counterexample", i.e. an L-structure* $\mathfrak{A}$ *that is a model of* $\Sigma$ *but in which* $\varphi$ *does not hold (thus* $\mathfrak{A}$ *is a model of* $\Sigma \cup \{\neg\varphi\}$*).*

*Proof*: By Lemma 1.4.1, the condition $\Sigma \nvdash \varphi$ is equivalent to the condition that the set $\Sigma_1 := \Sigma \cup \{\neg\varphi\}$ is consistent. Thus we must show that every consistent set $\Sigma \subseteq \mathrm{Sent}(L)$ possesses a model. (Here we have replaced $\Sigma \cup \{\neg\varphi\}$ by $\Sigma$.)

So let $\Sigma \subseteq \mathrm{Sent}(L)$ be consistent. We apply first Theorem 1.4.2 and then Theorem 1.4.4 to this $\Sigma$. We obtain thereby a $\Sigma^* \subseteq \mathrm{Sent}(L')$ with properties (I) and (II) on page 33. Here, $L'$ is the extension language of $L$ constructed in Theorem 1.4.2. Let

$$A = \mathrm{CT}/\approx \tag{1.5.2.1}$$

be the set of equivalence classes of constant terms of the language $L'$, constructed in (1.4.5.4). Further, let $\mathscr{R}_i$ (for $i \in I$), $\mathscr{F}_j$ (for $j \in J$) and $\overline{c_k}$ (for $k \in K'$) be the interpretations of the relation symbols $R_i$, the function symbols $f_j$ and the constant symbols $c_k$, given in (1.4.5.7), (1.4.5.9) and (1.4.5.10), respectively. Then

$$\mathfrak{A} = \langle A;\ (\mathscr{R}_i)_{i\in I};\ (\mathscr{F}_j)_{j\in J};\ (c_k)_{k\in K'}\rangle \tag{1.5.2.2}$$

is, finally, an $L'$-structure. By definition,

$$\mathscr{R}_i = R_i^{\mathfrak{A}}, \qquad \mathscr{F}_j = f_j^{\mathfrak{A}}, \qquad \overline{c_k} = c_k^{\mathfrak{A}}. \tag{1.5.2.3}$$

From our special definition of functions, we even have

$$t^{\mathfrak{A}}\,[h] = \overline{t} \tag{1.5.2.4}$$

for all $t \in \mathrm{CT}$ and all evaluations $h$ in $\mathfrak{A}$. Indeed, by induction on the construction of a constant term we have

$$\begin{aligned} f_j(t_1,\ldots,t_{\mu(j)})^{\mathfrak{A}}\,[h] &= f_j^{\mathfrak{A}}\left(t_1^{\mathfrak{A}}\,[h],\ldots,t_{\mu(j)}^{\mathfrak{A}}\,[h]\right) && \text{by (1.5.0.4)}\\ &= \mathscr{F}_j(\overline{t_1},\ldots,\overline{t_{\mu(j)}}) && \text{by (1.5.2.3) and ind. hyp.}\\ &= \overline{f_j(t_1,\ldots,t_{\mu(j)})} && \text{by (1.4.5.9).}\end{aligned}$$

Now we show that for every $\varphi \in \mathrm{Sent}(L')$ and every evaluation $h$ in $\mathfrak{A}$,

$$\mathfrak{A} \models \varphi\,[h] \quad \text{iff} \quad \varphi \in \Sigma^*. \tag{1.5.2.5}$$

Then, in particular, $\mathfrak{A}$ will be a model of $\Sigma^*$. We shall prove (1.5.2.5) by induction on the construction of $\varphi$ – more precisely, by induction on the number of logical symbols $\neg$, $\wedge$ and $\forall$ used in the construction of $\varphi$.

If this number is 0, then we are dealing with an atomic formula. But in this case the definitions furnish us with the required equivalence, since for constant terms $t_1, t_2, \ldots$ we have, first,

$$\begin{array}{lll}
\mathfrak{A} \models t_1 \doteq t_2\,[h] & \text{iff } t_1^{\mathfrak{A}}\,[h] = t_2^{\mathfrak{A}}\,[h] & \text{by (1.5.0.5)} \\
& \text{iff } \overline{t_1} = \overline{t_2} & \text{by (1.5.2.4)} \\
& \text{iff } t_1 \doteq t_2 \in \Sigma^* & \text{by (1.4.5.6), (1.4.5.3),} \\
& & \text{and (1.4.6),}
\end{array}$$

and, second,

$$\begin{array}{lll}
\mathfrak{A} \models R_i(t_1,\ldots)\,[h] & \text{iff } R_i^{\mathfrak{A}}(t_1^{\mathfrak{A}}\,[h],\ldots) & \text{by (1.5.0.6)} \\
& \text{iff } \mathscr{R}_i(\overline{t_1},\ldots) & \text{by (1.5.2.3) and (1.5.2.4)} \\
& \text{iff } R_i(t_1,\ldots) \in \Sigma^* & \text{by (1.4.5.7) and (1.4.6).}
\end{array}$$

Next, if the sentence $\varphi$ is of the form $\neg\alpha$ or $(\alpha \wedge \beta)$, then $\alpha$ and $\beta$ are likewise sentences. We have:

$$\begin{array}{lll}
\mathfrak{A} \models \neg\alpha\,[h] & \text{iff } \mathfrak{A} \not\models \alpha\,[h] & \text{by (1.5.0.7)} \\
& \text{iff } \alpha \notin \Sigma^* & \text{by ind. hyp. (1.5.2.5)} \\
& \text{iff } \neg\alpha \in \Sigma^* & \text{by (1.4.7)(a), and}
\end{array}$$

$$\begin{array}{lll}
\mathfrak{A} \models (\alpha \wedge \beta)\,[h] & \text{iff } (\mathfrak{A} \models \alpha\,[h] \text{ and } \mathfrak{A} \models \beta\,[h]) & \text{by (1.5.0.8)} \\
& \text{iff } (\alpha \in \Sigma^* \text{ and } \beta \in \Sigma^*) & \text{by ind. hyp. (1.5.2.5)} \\
& \text{iff } (\alpha \wedge \beta) \in \Sigma^* & \text{by (1.4.7)(b).}
\end{array}$$

Finally, if the sentence $\varphi$ is of the form $\forall x\,\psi$, then for $t \in \mathrm{CT}$, $\psi(x/t)$ is obviously again a sentence, since $\mathrm{Fr}(\psi) \subseteq \{x\}$. And $\psi(x/t)$ is more simply built than $\varphi$, as far as the number of logical symbols. Therefore, using Lemma 1.5.3, proved just below, we have:

$$\begin{array}{llll}
\mathfrak{A} \models \forall x\,\psi\,[h] & \text{iff } \mathfrak{A} \models \psi\,[h\binom{x}{a}] & \text{for all } a \in A & \text{by (1.5.0.9)} \\
& \text{iff } \mathfrak{A} \models \psi\,[h\binom{x}{\overline{t}}] & \text{for all } t \in \mathrm{CT} & \text{by (1.5.2.1) \& (1.4.5.5)} \\
& \text{iff } \mathfrak{A} \models \psi(x/t)\,[h] & \text{for all } t \in \mathrm{CT} & \text{by (1.5.3) for } L' \qquad (1.5.2.6) \\
& \text{iff } \psi(x/t) \in \Sigma^* & \text{for all } t \in \mathrm{CT} & \text{by ind. hyp. (1.5.2.5)} \\
& \text{iff } \forall x\,\psi \in \Sigma^* & & \text{by (1.4.7)(c).}
\end{array}$$

We have proved, finally (modulo Lemma 1.5.3 below), that all $L'$-sentences $\varphi \in \Sigma^*$ (and only those) hold in the $L'$-structure $\mathfrak{A}$ (1.5.2.2). It is therefore clear that all $L$-sentences $\varphi \in \Sigma$ hold in the $L$-structure

$$\mathfrak{A}|_L := \langle A;\ (\mathscr{R}_i)_{i\in I};\ (\mathscr{F}_j)_{j\in J};\ (c_k)_{k\in K}\rangle.$$

Thus, $\Sigma$ possesses a model. □

The following lemma (used in (1.5.2.6) above) is of a technical nature.

**Lemma 1.5.3.** *Let $\mathfrak{A}$ be L-structure, h an evaluation in $\mathfrak{A}$, $\varphi$ an L-formula, x a variable and t an L-term that is free for x in $\varphi$. Then, writing $a = t^{\mathfrak{A}}[h]$ (1.5.0.4), we have:*

$$\mathfrak{A} \models \varphi\left[h\binom{x}{a}\right] \quad \textit{iff} \quad \mathfrak{A} \models \varphi(x/t)[h]. \tag{1.5.3.1}$$

*Proof*: For terms $t_1$ one shows easily, by induction on their construction, that

$$t_1^{\mathfrak{A}}\left[h\binom{x}{a}\right] = t_1(x/t)^{\mathfrak{A}}[h]. \tag{1.5.3.2}$$

Now we shall prove (1.5.3.1) by induction on the construction of the formula $\varphi$. In the atomic formula $t_1 \doteq t_2$, for example, we have:

$$\begin{aligned}
\mathfrak{A} \models t_1 \doteq t_2 \left[h\binom{x}{a}\right] \quad &\text{iff} \quad t_1^{\mathfrak{A}}\left[h\binom{x}{a}\right] = t_2^{\mathfrak{A}}\left[h\binom{x}{a}\right] && \text{by (1.5.0.5)}\\
&\text{iff} \quad t_1(x/t)^{\mathfrak{A}}[h] = t_2(x/t)^{\mathfrak{A}}[h] && \text{by (1.5.3.2)}\\
&\text{iff} \quad \mathfrak{A} \models (t_1 \doteq t_2)\,(x/t)\,[h] && \text{by (1.5.0.5).}
\end{aligned} \tag{1.5.3.3}$$

In (1.5.3.3) we have enclosed $t_1 \doteq t_2$ in parentheses in order to indicate the scope of application of the syntactic operation $(x/t)$. (Note that the notation $(x/t)$ (1.2.0.8) belongs not to the object language, but to the metalanguage!) It is clear that $(t_1 \doteq t_2)\,(x/t)$ is $t_1\,(x/t) \doteq t_2\,(x/t)$.

The other case (1.5.0.6) of atomic formulae $\varphi$ is handled analogously.

The cases in which $\varphi$ is of the form $\neg\varphi$ or $(\alpha \wedge \beta)$ are likewise easily handled, by referring back to the components $\alpha$, or $\alpha$ and $\beta$, respectively.

There remains only the case where $\varphi$ is of the form $\forall y\,\psi$. Here we distinguish two (sub)cases:

*Case 1*: Either $y$ is $x$, or $x \notin \mathrm{Fr}(\psi)$. Under either of these assumptions, $x \notin \mathrm{Fr}(\forall y\,\psi)$. Therefore, using Lemma 1.5.1,

$$\mathfrak{A} \models \forall y\,\psi\left[h\binom{x}{a}\right] \quad \text{iff} \quad \mathfrak{A} \models \forall y\,\psi\,[h].$$

This, however, is (1.5.3.1), since $(\forall y\,\psi)\,(x/t)$ is clearly $\forall y\,\psi$.

*Case 2*: $y$ is not $x$, and $x \in \mathrm{Fr}(\psi)$. In this case $y$ cannot occur in $t$, since, by hypothesis, $t$ is free for $x$ in $\forall y\,\psi$ (p. 12). It follows, using Lemma 1.5.1(a), that for every $a' \in |\mathfrak{A}|$,

$$a = t^{\mathfrak{A}}[h] = t^{\mathfrak{A}}\left[h\binom{y}{a'}\right]. \tag{1.5.3.4}$$

Writing $h' = h\binom{y}{a'}$, we then have

$$
\begin{array}{llll}
\mathfrak{A} \models \forall y\, \psi\left[h\binom{x}{a}\right] & \text{iff } \mathfrak{A} \models \psi\left[h\binom{x}{a}\binom{y}{a'}\right] & \text{for all } a' \in |\mathfrak{A}| & \text{by (1.5.0.9)} \\
& \text{iff } \mathfrak{A} \models \psi\left[h'\binom{x}{a}\right] & \text{for all } a' \in |\mathfrak{A}| & \text{since } y \text{ is not } x \\
& \text{iff } \mathfrak{A} \models \psi\left[h'\binom{x}{t^{\mathfrak{A}}[h']}\right] & \text{for all } a' \in |\mathfrak{A}| & \text{by (1.5.3.4)} \\
& \text{iff } \mathfrak{A} \models \psi(x/t)\,[h'] & \text{for all } a' \in |\mathfrak{A}| & \text{ind. hyp. (1.5.3.1) for } h' \\
& \text{iff } \mathfrak{A} \models \psi(x/t)\left[h\binom{y}{a'}\right] & \text{for all } a' \in |\mathfrak{A}| & \text{def. of } h' \\
& \text{iff } \mathfrak{A} \models \forall y\, \psi(x/t)\,[h] & & \text{by (1.5.0.9)} \\
& \text{iff } \mathfrak{A} \models (\forall y\, \psi)\,(x/t)\,[h] & & \text{since } y \text{ is not } x. \quad \square
\end{array}
$$

The next theorem assures us of the *correctness* or *soundness* of the concept of proof developed in Section 1.3. It asserts that everything that can be proved from an axiom system $\Sigma$ also holds in every model of $\Sigma$. Thus, if we have a model of $\Sigma \cup \{\neg\varphi\}$, then obviously $\varphi$ cannot be proved from $\Sigma$. Considered formally, this is the converse of the implication asserted in Gödel's Completeness Theorem (1.5.2).

**Theorem 1.5.4** (Soundness Theorem). *Suppose* $\Sigma \subseteq \mathrm{Sent}(L)$, $\varphi \in \mathrm{Sent}(L)$ *and* $\Sigma \vdash \varphi$. *Then* $\varphi$ *holds in every model of* $\Sigma$.

*Proof*: Let $\mathfrak{A}$ be a model of $\Sigma$, and $\varphi_1, \ldots, \varphi_n$ be a proof from $\Sigma$. We shall show that $\mathfrak{A} \models \varphi_i$ for $i \in \{1, 2, \ldots, n\}$, by induction on $i$. This will, in particular, prove the theorem.

*Basis step of the induction*: Let $i = 1$. Then either $\varphi_i \in \Sigma$ (1.3.0.1), or $\varphi_i$ is a logical axiom (1.3.0.2). In the first case, $\varphi_i$ holds in $\mathfrak{A}$ by hypothesis.

In the second case, suppose, first, that $\varphi_i$ is an instance of a tautological form $\Phi$ in the sentential variables $A_0, \ldots, A_m$, arising by the replacement of $A_j$ by the $L$-formula $\psi_j$, for $0 \leq j \leq m$ (p. 16). Let $h$ be any evaluation in $\mathfrak{A}$. Define a truth assignment $\mathscr{H}$ on the variables $A_0, \ldots, A_m$ as follows:

$$\mathscr{H}(A_j) = \begin{cases} T & \text{if } \mathfrak{A} \models \psi_j\,[h], \\ F & \text{if } \mathfrak{A} \not\models \psi_j\,[h]. \end{cases}$$

Then $\mathscr{H}(\Phi) = T$ iff $\mathfrak{A} \models \varphi_i\,[h]$, since the equations (1.3.0.4) and (1.3.0.5) giving the recursive definition of $\mathscr{H}$ on arbitrary sentential forms $\Phi$ agree, respectively, with the equations (1.5.0.7) and (1.5.0.8) giving (the relevant part of) the recursive definition of $\mathfrak{A} \models \varphi_i\,[h]$. But $\mathscr{H}(\Phi) = T$, since $\Phi$ is tautological; therefore $\mathfrak{A} \models \varphi_i\,[h]$. Since $h$ was arbitrary, we conclude that $\mathfrak{A} \models \varphi_i$ (p. 40).

In the case where $\varphi_i$ is one of the equality axioms (1.3.0.9), one may convince oneself just as easily that $\mathfrak{A} \models \varphi_i$.

There remain the cases where $\varphi_i$ is an instance of the quantifier axioms (A1) (1.3.0.7) or (A2) (1.3.0.8).

First let $\varphi_i$ be of the form

$$\forall x(\alpha \to \beta) \to (\alpha \to \forall x \beta),$$

where $x$ is not free in $\alpha$. According to the definition of satisfaction (p. 40), we must show, for every evaluation $h$, that:

the hypothesis

$$\text{for all } a \in |\mathfrak{A}|,\ \mathfrak{A} \models (\alpha \to \beta)\left[h\binom{x}{a}\right] \tag{1.5.4.1}$$

implies

$$\text{if } \mathfrak{A} \models \alpha\,[h], \text{ then for all } a \in |\mathfrak{A}|,\ \mathfrak{A} \models \beta\left[h\binom{x}{a}\right]. \tag{1.5.4.2}$$

So assume (1.5.4.1), and suppose $\mathfrak{A} \models \alpha\,[h]$ and $a \in |\mathfrak{A}|$. Then $\mathfrak{A} \models \alpha\left[h\binom{x}{a}\right]$ by Lemma 1.5.1(b), since $x \notin \mathrm{Fr}(\alpha)$. (1.5.4.1) and (1.5.0.10) then give $\mathfrak{A} \models \beta\left[h\binom{x}{a}\right]$, proving (1.5.4.2), as required in this "(A2)" case.

Second, let $\varphi_i$ be of the form

$$\forall x\,\alpha \to \alpha(x/t),$$

where $t$ is free for $x$ in $\alpha$. For each evaluation $h$ we must show that: the hypothesis

$$\mathfrak{A} \models \alpha\left[h\binom{x}{a}\right] \text{ for all } a \in |\mathfrak{A}| \tag{1.5.4.3}$$

implies

$$\mathfrak{A} \models \alpha(x/t)\,[h]. \tag{1.5.4.4}$$

By Lemma 1.5.3, (1.5.4.4) is equivalent to $\mathfrak{A} \models \alpha\left[h\binom{x}{a}\right]$ with $a = t^{\mathfrak{A}}\,[h]$. But this is a special case of (1.5.4.3).

*Induction step*: Assume, for all $j < i$, that we have already proved $\mathfrak{A} \models \varphi_j$. We must show $\mathfrak{A} \models \varphi_i$.

If $\varphi_i$ is a member of $\Sigma$ or a logical axiom, then we obtain $\mathfrak{A} \models \varphi_i$ as in the basis step of the induction, above. If $\varphi_i$ comes about by means of (MP), then there are $j, k < i$ with $\varphi_k$ being $\varphi_j \to \varphi_i$. For each evaluation $h$ we then have $\mathfrak{A} \models (\varphi_j \to \varphi_i)\,[h]$ and $\mathfrak{A} \models \varphi_j\,[h]$. This immediately gives $\mathfrak{A} \models \varphi_i\,[h]$, using (1.5.0.10). Since $h$ was arbitrary, $\mathfrak{A} \models \varphi_i$.

Finally, if $\varphi_i$ comes about by means of $(\forall)$, then there exists a $j < i$ and a variable $x$ such that $\varphi_i$ is $\forall x\,\varphi_j$. The inductive hypothesis is that $\mathfrak{A} \models \varphi_j$. From this follows $\mathfrak{A} \models \forall x\,\varphi_j$ (1.5.1.4). □

**Corollary 1.5.5.** *A set $\Sigma \subseteq \mathrm{Sent}(L)$ is consistent if and only if it possesses a model.*

*Proof*: If $\Sigma$ is consistent, then $\Sigma$ possesses a model by Gödel's Completeness Theorem 1.5.2. Conversely, if $\Sigma$ possesses a model, then no contradiction can be deduced from $\Sigma$, according to Theorem 1.5.4. □

From this corollary, which actually summarizes Theorems 1.5.2 and 1.5.4, we obtain the most important theorem of model theory (later named the Compactness Theorem):

**Theorem 1.5.6** (Finiteness Theorem). *A set $\Sigma \subseteq \mathrm{Sent}(L)$ possesses a model if and only if every finite subset $\Pi$ of $\Sigma$ possesses a model.*

*Proof*: If $\Sigma$ possesses a model, then every finite subset of $\Sigma$ possesses the same model. It remains to show the converse. Assume that $\Sigma$ possesses no model. Then $\Sigma$ would be inconsistent, by (1.5.5). However, since only finitely many elements of $\Sigma$ can occur in any proof of any contradiction from $\Sigma$, some finite subset $\Pi \subseteq \Sigma$ would already be inconsistent. By (1.5.5) again, $\Pi$ would have no model. This proves the converse. □

*Remark 1.5.7* (Model theoretic proofs of the Finiteness Theorem). The above proof of the Finiteness Theorem made essential use of Gödel's Completeness Theorem, and hence of the concept of proof. In the above argument we see immediately how the finiteness of the concept of proof has an impact. Observe, however, that the *statement* of the Finiteness Theorem makes no reference to the concept of proof. In fact, there are other proofs of this theorem, which are purely model theoretic. They can be expressed using only the concepts of "formal language" and "model." In Section 2.6 we shall carry out such a proof, by means of ultraproducts. For a deeper understanding of the Finiteness Theorem, however, the proof using Gödel's Completeness Theorem seems to us to be better.

## 1.6 Axiomatization of Several Mathematical Theories

In this section we wish to axiomatize several mathematical theories in the framework of the formal languages introduced here. In particular, we shall give an axiom system (in a suitable language) for set theory. First, however, we shall clarify what we mean by a mathematical theory.

The word "theory" in mathematics has many uses, and cannot easily be defined comprehensively. Consider "number theory", as an example: what can be said with certainty is that in this "theory" one investigates the set $\mathbb{N}$ of natural numbers (or also the set $\mathbb{Z}$ of integers) and the properties of the operations "addition" and "multiplication" defined on those numbers. Usually one also associates with a certain mathematical "theory", implicitly or explicitly, the methods that are applied in it; for example, one speaks of "analytic number theory" or "algebraic number theory" or even "modern number theory". In the case of the attribute "modern", other or new methods are not necessarily meant; often it is just a matter of a representation in a new, "modern" language. All these methods of investigation have one goal in common:

*One wishes to produce, as far as possible, all sentences true in $\mathbb{N}$ (or in $\mathbb{Z}$).*

Analogously, in "group theory" one wants to produce, as far as possible, all sentences true in a specified class of groups. We take this goal as the motivation for the following definition.

Let $L = (\lambda, \mu, K)$ be a formal language and $\mathscr{M}$ a nonempty class of $L$-structures. Then we define *the $L$-theory of $\mathscr{M}$* to be the set

$$\mathrm{Th}(\mathscr{M}) := \{\alpha \in \mathrm{Sent}(L) \mid \mathfrak{A} \models \alpha, \text{ for all } \mathfrak{A} \in \mathscr{M}\}. \tag{1.6.0.1}$$

Thus, $\mathrm{Th}(\mathscr{M})$ consists exactly of all $L$-sentences that hold in all structures in $\mathscr{M}$. The set $\mathrm{Th}(\mathscr{M})$ of sentences possesses two important properties:

(1) $\mathrm{Th}(\mathscr{M})$ is consistent, and

(2) $\mathrm{Th}(\mathscr{M})$ is deductively closed.

Here, a subset $\Sigma \subseteq \mathrm{Sent}(L)$ is called *deductively closed* (cf. Lemma 1.4.6) if, for every $L$-sentence $\alpha$, $\Sigma \vdash \alpha$ implies $\alpha \in \Sigma$. Property (1) follows from Corollary 1.5.5 and the nonemptiness of $\mathscr{M}$. Property (2) is derived as follows: Suppose $\mathrm{Th}(\mathscr{M}) \vdash \alpha$. Since every $\mathfrak{A} \in \mathscr{M}$ is obviously a model of $\mathrm{Th}(\mathscr{M})$, $\mathfrak{A}$ is also a model of $\alpha$, by Theorem 1.5.4. Therefore $\alpha$ belongs to $\mathrm{Th}(\mathscr{M})$, by (1.6.0.1).

More generally, we call a subset $T$ of $\mathrm{Sent}(L)$ an *$L$-theory* if $T$ is consistent and deductively closed. Every $L$-theory in this sense is actually *the* $L$-theory of a class of $L$-structures, namely, just the class $\mathscr{M}$ of all models of $T$, i.e. the class

$$\mathscr{M} = \{\mathfrak{A} \mid \mathfrak{A} \text{ is an } L\text{-structure and } \mathfrak{A} \models T\}.$$

This is easy to see: by (1.6.0.1), $T \subseteq \mathrm{Th}(\mathscr{M})$ holds trivially, since each $\alpha \in T$ naturally holds in all models of $T$. If, conversely, $\alpha$ holds in all models of $T$, then the Completeness Theorem 1.5.2 immediately yields $T \vdash \alpha$. Therefore $\alpha \in T$, by the deductive closedness of $T$.

While for any given nonempty class $\mathscr{M}$ of $L$-structures we can, indeed, define the set $T = \mathrm{Th}(\mathscr{M})$ purely abstractly, as a rule this set will be completely intractable. We shall return to this problem in Appendix A. Sometimes, however, it is possible to give a reasonable system $\Sigma$ of axioms for $T$. Here we call $\Sigma$ an *axiom system* for $T$ if

$$T = \{\alpha \in \mathrm{Sent}(L) \mid \Sigma \vdash \alpha\}.$$

Observe that the set $\{\alpha \in \mathrm{Sent}(L) \mid \Sigma \vdash \alpha\}$, which we wish to denote also by $\mathrm{Ded}(\Sigma)$, is deductively closed. Indeed, if $\alpha_1, \ldots, \alpha_n \in \mathrm{Ded}(\Sigma)$ and $\{\alpha_1, \ldots, \alpha_n\} \vdash \alpha$, then we have $n$ proofs of $\alpha_1, \ldots, \alpha_n$ from $\Sigma$, and a proof of $\alpha$ from $\{\alpha_1, \ldots, \alpha_n\}$; from these $n+1$ proofs we can easily assemble a proof of $\alpha$ directly from $\Sigma$. By a "reasonable" axiom system $\Sigma$ for $T$ we mean one that is effectively enumerable, e.g. finite. Appendix A will explain more precisely how the concept of "effective enumerability" can be given a definition. In the following examples we shall write down the corresponding axiom systems concretely.

Usually the most interesting case of a possible axiomatization of the $L$-theory of a class $\mathscr{M}$ is that in which $\mathscr{M}$ has exactly one $L$-structure $\mathfrak{A}$. In this case we also write $\mathrm{Th}(\mathfrak{A})$ for $\mathrm{Th}(\mathscr{M})$, and (1.6.0.1) simplifies to

$$\mathrm{Th}(\mathfrak{A}) = \{\alpha \in \mathrm{Sent}(L) \mid \mathfrak{A} \models \alpha\}.$$

Here we also speak of the "$L$-theory of $\mathfrak{A}$". For example, if

$$\mathfrak{N} = \langle \mathbb{N}; +^{\mathbb{N}}; \cdot^{\mathbb{N}} \rangle,$$

then $\mathrm{Th}(\mathfrak{N})$ is what we want to be understood by the phrase "number theory", namely, simply the set of all sentences (of the formal language $L$ appropriate for $\mathfrak{A}$) that are true in $\mathfrak{N}$.

For an $L$-structure $\mathfrak{A}$, the $L$-theory $T := \mathrm{Th}(\mathfrak{A})$ has an excellent property: since for every $L$-sentence $\alpha$,

$$\mathfrak{A} \models \alpha \quad \text{or} \quad \mathfrak{A} \models \neg\alpha,$$

we obtain, correspondingly,

$$\alpha \in T \quad \text{or} \quad \neg\alpha \in T.$$

We call an arbitrary $L$-theory $T$ with this property *complete*. More generally we want to call an arbitrary set $\Sigma$ of $L$-sentences *complete* if, for every $L$-sentence $\alpha$,

$$\Sigma \vdash \alpha \quad \text{or} \quad \Sigma \vdash \neg\alpha. \tag{1.6.0.2}$$

The latter definition does not clash with the former in the case where $\Sigma$ itself is already an $L$-theory, since in that case $\Sigma$ would be deductively closed. And the latter definition is more general, as one sees by considering either the consistency, or the deductive closedness, of an $L$-theory $T$. If $\Sigma$ is consistent and $T = \mathrm{Ded}(\Sigma)$, then clearly $T$ is complete in the first sense if and only if $\Sigma$ is in the second sense.

With this we have finally arrived at a purely syntactical concept, which is then also meaningful when one adopts the finitist standpoint. If $L$ is a language permitting us to express all reasonable mathematical concepts, and if some set $\Sigma$ is a concrete, consistent axiom system in $L$ (we shall, further below, present the Zermelo–Fraenkel axiom system for set theory as one such system), then the completeness of $\Sigma$ would mean that one could prove (from $\Sigma$) every "true" mathematical sentence. Then the truth of a sentence would be nothing other than its provability. In Appendix A we explain that such an axiom system cannot exist. There we shall see that, for example, the Zermelo–Fraenkel axiom system is incomplete, i.e. that there exist sentences $\alpha$ that are neither provable nor refutable from it (where "$\alpha$ is refutable" means that $\neg\alpha$ is provable).

Just below we shall introduce a series of axiom systems whose completeness we shall prove in Chapter 3. If $\Sigma \subseteq \mathrm{Sent}(L)$ is complete, and $\mathfrak{A}$ is a model of $\Sigma$, then, in particular,

$$\mathrm{Th}(\mathfrak{A}) = \mathrm{Ded}(\Sigma), \tag{1.6.0.3}$$

i.e. $\mathrm{Ded}(\Sigma)$ is the theory of the $L$-structure $\mathfrak{A}$. Indeed, since $\mathfrak{A}$ is a model of $\Sigma$, we have, on the one hand, $\Sigma \subseteq \mathrm{Th}(\mathfrak{A})$ and hence $\mathrm{Ded}(\Sigma) \subseteq \mathrm{Th}(\mathfrak{A})$. On the other hand, both $T_1 := \mathrm{Ded}(\Sigma)$ and $T_2 := \mathrm{Th}(\mathfrak{A})$ are, first, $L$-theories (for $T_1$ this requires noticing that $\Sigma$ is consistent), and second, complete; and whenever we have an inclusion

$T_1 \subseteq T_2$ between two complete $L$-theories, we even have equality. Indeed, if $\alpha \in T_2$ and $\alpha \notin T_1$, then $\neg\alpha \in T_1$ and thence $\neg\alpha \in T_2$, contradicting the consistency of $T_2$. We record this as a Lemma:

**Lemma 1.6.1.** *If $T_1$ and $T_2$ are complete L-theories with $T_1 \subseteq T_2$, then $T_1 = T_2$.* □

Now we want to introduce a series of $L$-theories; for each one, we shall, either in this section or in Chapters 3 or 4, investigate whether it is complete. Each theory considered here will be presented as the deductive closure of a concrete axiom system. While setting up this axiom system, we shall usually have a particular $L$-structure $\mathfrak{A}$ in view. We usually begin with several general properties of $\mathfrak{A}$, which we make into axioms, and then we try, by a systematic enlargement of this system, to arrive, finally, at an axiomatization of $\mathrm{Th}(\mathfrak{A})$, i.e. at a complete axiom system.

*1. Dense linear orderings with no extrema*

The first structure whose theory we want to axiomatize is

$$\mathfrak{R} = \langle \mathbb{R}; <^{\mathbb{R}} \rangle.$$

This is a structure in the language $L$, for which $J = \emptyset$, $K = \emptyset$ and $I$ consists of a single element $i$ with $\lambda(i) = 2$ (recall (1.2.0.10)–(1.2.0.11)). For the relation symbol $R_i$ we write $<$. The following are $L$-sentences that hold in $\mathfrak{R}$:

$$
\begin{array}{lll}
\mathrm{O}_1\colon & \forall x & \neg\, x < x \\
\mathrm{O}_2\colon & \forall x, y, z & (x < y \wedge y < z \rightarrow x < z) \\
\mathrm{O}_3\colon & \forall x, y & (x < y \vee x \doteq y \vee y < x) \\
\mathrm{O}_4\colon & \forall x, y & (x < y \rightarrow \exists z\,(x < z \wedge z < y)) \\
\mathrm{O}_5\colon & \forall x\, \exists y, z & (y < x \wedge x < z)
\end{array}
$$

$\mathrm{O}_1$ and $\mathrm{O}_2$ are called the axioms of a *partial ordering*; $\mathrm{O}_1$–$\mathrm{O}_3$ are the axioms of a *linear ordering*. The addition of $\mathrm{O}_4$ says that this linear ordering is *dense*; and the addition of $\mathrm{O}_5$ says that it possesses no *extrema*.

In Section 3.2 we shall see that $\mathrm{O}_1$–$\mathrm{O}_5$ is complete. $\mathrm{O}_1$–$\mathrm{O}_4$ is not complete, since, for example, the sentence $\mathrm{O}_5$ is true in $\mathfrak{R}$, but not in $\mathfrak{R}^+ := \langle \mathbb{R}^+; <^{\mathbb{R}^+} \rangle$, where $\mathbb{R}^+$ denotes the set of nonnegative real numbers.

*2. Torsion-free, divisible, Abelian groups*

Next, let us consider the structure

$$\mathfrak{R} = \langle \mathbb{R}; +^{\mathbb{R}}; 0^{\mathbb{R}} \rangle,$$

which we have, for the sake of simplicity, again denoted by $\mathfrak{R}$. This time, $L$ has a binary function symbol $+$ and the constant symbol 0. The following $L$-sentences hold in $\mathfrak{R}$:

$$\begin{aligned}
&\mathrm{G}_1: && \forall x,y,z(x+y)+z \doteq x+(y+z)\\
&\mathrm{G}_2: && \forall x \quad x+0 \doteq x\\
&\mathrm{G}_3: && \forall x\, \exists y\; x+y \doteq 0\\
&\mathrm{G}_4: && \forall x,y \;\; x+y \doteq y+x\\
&\mathrm{G}_{5,n}: && \forall x \quad (nx \doteq 0 \rightarrow x \doteq 0)\\
&\mathrm{G}_{6,n}: && \forall x\, \exists y\; ny \doteq x
\end{aligned}$$

Here, for each natural number $n \geq 1$, the (informal) expression $nx$ abbreviates the term $x+x+\cdots+x$, the $n$-fold sum of $x$.[4] $\mathrm{G}_1$–$\mathrm{G}_3$ are usually called the *group axioms*; $\mathrm{G}_4$ says that this group is *Abelian*. The set $\{\, \mathrm{G}_{5,n} \mid n \geq 1 \,\}$ of axioms expresses the *torsion free* property of a group; the set $\{\, \mathrm{G}_{6,n} \mid n \geq 1 \,\}$ expresses its *divisibility*. In Section 3.2 we shall prove the completeness of the axiom system

$$\{\mathrm{G}_1, \mathrm{G}_2, \mathrm{G}_3, \mathrm{G}_4\} \cup \{\, \mathrm{G}_{5,n} \mid n \geq 1 \,\} \cup \{\, \mathrm{G}_{6,n} \mid n \geq 1 \,\} \cup \{\exists x\, x \neq 0\}$$

of nontrivial, torsion-free, divisible, Abelian groups.

*3. Ordered, divisible, Abelian groups*

If we consider both the additive group structure and the ordering on $\mathbb{R}$ simultaneously, then one arrives, in the correspondingly extended language, at the structure

$$\mathfrak{R} = \langle \mathbb{R};\ <^{\mathbb{R}};\ +^{\mathbb{R}};\ 0^{\mathbb{R}} \rangle.$$

This is an *ordered, divisible, Abelian group*, i.e. it is a model of all of the above-listed O-axioms, all G-axioms, and additionally the axiom

$$\mathrm{OG}: \quad \forall x,y,z\, (x < y \;\rightarrow\; x+z < y+z).$$

The totality of all these axioms is, as we shall show in Chapter 4, complete. One sees easily that here the axioms $\mathrm{O}_4$ and $\mathrm{G}_{5,n}$ are superfluous, and $\mathrm{O}_5$ can be replaced by $\exists x\, x \neq 0$. We shall also show that the axiom system

$$\{\mathrm{O}_1, \mathrm{O}_2, \mathrm{O}_3, \mathrm{G}_1, \mathrm{G}_2, \mathrm{G}_3, \mathrm{G}_4, \mathrm{OG}\} \cup \{\, \mathrm{G}_{6,n} \mid n \geq 1 \,\} \cup \{\exists x\, x \neq 0\}$$

is complete. Models of the first part of the above union are called *ordered Abelian groups*.

[4] More precisely, this term should be written as $(\cdots((x+x)+x)+\cdots)+x$, since the arity of $+$ is 2. No ambiguity will result from dropping these parentheses, thanks to $\mathrm{G}_1$.

*4. Discrete ordered Abelian groups*

Another ordered Abelian group whose theory interests us is

$$\mathfrak{Z} = \langle \mathbb{Z};\ <^{\mathbb{Z}};\ +^{\mathbb{Z}};\ 0^{\mathbb{Z}} \rangle.$$

In order to axiomatize its theory, we observe, first, that it is not divisible: it possesses a smallest, positive element, namely the natural number 1. An ordered Abelian group, which, like $\mathfrak{Z}$, satisfies the axiom

$$\text{DO:}\quad \exists x\,(0 < x \wedge \forall y\,(0 < y \rightarrow x \doteq y \vee x < y))$$

is called *discrete ordered*. The element $x$ whose existence DO asserts is unique, by $O_1$ and $O_2$.

*5. $\mathbb{Z}$-groups*

To obtain a complete axiomatization of $\mathfrak{Z}$, it is advisable to add to the language of ordered groups a constant symbol for the smallest positive element. Let this constant symbol be 1. Thus, we now consider the structure

$$\mathfrak{Z} = \langle \mathbb{Z};\ <^{\mathbb{Z}};\ +^{\mathbb{Z}};\ 0^{\mathbb{Z}},\ 1^{\mathbb{Z}} \rangle.$$

Besides the axioms for discrete ordered Abelian groups, $\mathfrak{Z}$ satisfies the following family of axioms:

$$\text{D}_n:\quad \forall x \exists y \left( \bigvee_{\nu=1}^{n} x + \nu 1 \doteq ny \right). \qquad (1.6.1.1)$$

Here $n = 1, 2, \ldots$, and for $1 \leq \nu \leq n$, the (informal) expression $\nu 1$ abbreviates the term $1 + 1 + \cdots + 1$, the $n$-fold sum of 1. As in (1.2.0.12), the (informal) expression $\left(\bigvee_{\nu=1}^{n} \varphi_\nu\right)$ is an abbreviation for the formula

$$(\varphi_1 \vee \cdots \vee \varphi_n).$$

A discrete ordered Abelian group that is, in addition, a model of $\{D_n \mid n \geq 1\}$ is called a $\mathbb{Z}$-*group*. The reason for this name is the fact (proven later, in Chapter 4) that the axiom system for $\mathbb{Z}$-groups is complete, and $\mathfrak{Z}$ is a model of it.

*6. Ordered fields, real closed fields*

Next we consider the structure

$$\mathfrak{R} = \langle \mathbb{R};\ +^{\mathbb{R}},\ -^{\mathbb{R}},\ \cdot^{\mathbb{R}};\ 0^{\mathbb{R}},\ 1^{\mathbb{R}} \rangle,$$

which we wish again to denote by the letter $\mathfrak{R}$, even though this structure extends previous structures by the same name. Now $\mathfrak{R}$ satisfies the following axioms:

the *field axioms:*

$$\begin{aligned}
&K_1: && 0 \neq 1 && (1.6.1.2)\\
&K_2: && \forall x,y,z \quad x+(y+z) \doteq (x+y)+z\\
&K_3: && \forall x \quad x+0 \doteq x\\
&K_4: && \forall x \quad x+(-x) \doteq 0\\
&K_5: && \forall x,y,z \quad x\cdot(y\cdot z) \doteq (x\cdot y)\cdot z\\
&K_6: && \forall x \quad x\cdot 1 \doteq x\\
&K_7: && \forall x \exists y \quad (x \doteq 0 \vee x\cdot y \doteq 1)\\
&K_8: && \forall x,y \quad x\cdot y \doteq y\cdot x\\
&K_9: && \forall x,y,z \quad (x+y)\cdot z \doteq x\cdot z + y\cdot z;
\end{aligned}$$

the order axioms $O_1$, $O_2$, $O_3$, together with:

$$\begin{aligned}
&OK_1: && \forall x,y,z \quad (x<y \rightarrow x+z<y+z)\\
&OK_2: && \forall x,y \quad (0<x \wedge 0<y \rightarrow 0<x\cdot y);
\end{aligned}$$

and finally, in addition,

$$\begin{aligned}
&RK_1: && \forall x\, \exists y \quad (x<0 \vee x \doteq y^2)\\
&RK_{2n}: && \forall x_0,x_1,..,x_{2n}\, \exists y \quad y^{2n+1} + x_{2n}\cdot y^{2n} + \cdots + x_1\cdot y + x_0 \doteq 0.
\end{aligned}$$

In $RK_1$ and $RK_{2n}$ (where $n = 1,2,\ldots$), the (informal) expression $y^m$, for $m = 1,2,\ldots$, abbreviates the term $y\cdot y\cdot\cdots\cdot y$, the $m$-fold product of $y$.[5] And in $RK_{2n}$ we utilize the convention in algebra that the operation $\cdot$ is performed before the operation $+$, unless parentheses indicate otherwise. A model of the axioms $K_1$–$K_9$, $O_1$–$O_3$, $OK_1$ and $OK_2$ is called an *ordered field*. If the axioms $RK_1$ and $RK_{2n}$ join in (for all $n \geq 1$), then one speaks of a *real closed field*. The reason for this name is the fact (to be proved in Section 4.2) that the theory of real closed fields is complete, and $\mathfrak{R}$ is a model of it. It is easy to see that $RK_1$ and $RK_{2n}$ hold in $\mathfrak{R}$, by reflecting on the fact that a polynomial with real coefficients that has a sign-change in $\mathbb{R}$ also has a zero in $\mathbb{R}$, by the intermediate-value theorem for continuous functions.

*7. Algebraically closed fields*

We wish to close out the round of structures for whose theories we explicitly give an axiom system, by considering the field

$$\mathfrak{C} = \langle \mathbb{C};\, +^{\mathbb{C}},\, -^{\mathbb{C}},\, \cdot^{\mathbb{C}};\, 0^{\mathbb{C}},\, 1^{\mathbb{C}} \rangle$$

[5] More precisely, this term should be written as $(\cdots((y\cdot y)\cdot y)\cdot\cdots)\cdot y$, since the arity of $\cdot$ is 2. No ambiguity will result from dropping these parentheses, thanks to $K_5$.

of complex numbers. $\mathfrak{C}$ satisfies not only the field axioms $K_1$–$K_9$, but also the family of axioms

$$\mathrm{AK}_n:\quad \forall x_0,x_1,\ldots,x_n\ \exists y\quad y^{n+1}+x_n\cdot y^n+\cdots+x_0 \doteq 0, \tag{1.6.1.3}$$

for $n = 1,2,\ldots$. The satisfaction of $\mathrm{AK}_n$ by $\mathfrak{C}$ is equivalent to the "Fundamental Theorem of Algebra", which asserts that every non-constant polynomial with coefficients in $\mathbb{C}$ has a zero in $\mathbb{C}$. Fields that satisfy $\mathrm{AK}_n$ for all $n \geq 1$ are called *algebraically closed fields*. The axiom system for algebraically closed fields is not yet complete. But, as we shall show in Chapter 3, it will become complete as soon as we specify the characteristic of the field, i.e. when we adjoin as further axioms either, for some single prime number $p$, the sentence

$$\mathrm{C}_p:\quad p1 \doteq 0 \tag{1.6.1.4}$$

(where $p1$ again denotes the $p$-fold sum $1+\cdots+1$), or the set

$$\{\neg \mathrm{C}_q \mid q \in \mathbb{N},\ q \text{ prime}\}.$$

In the latter case one obtains a complete axiomatization of the theory of $\mathfrak{C}$.

*8. The natural numbers*

For all hitherto considered structures $\mathfrak{A}$, we were able to give an axiom system $\Sigma \subseteq \mathrm{Sent}(L)$ that, first, axiomatized $\mathrm{Th}(\mathfrak{A})$ (i.e. had the property that $\mathrm{Ded}(\Sigma) = \mathrm{Th}(\mathfrak{A})$), and, second, was "decidable". This should mean that there is an effective procedure that allows us, for an arbitrary, given $L$-sentence $\alpha$, to decide whether $\alpha$ is an axiom in $\Sigma$ or not. A glance at the various axiom systems we have given will easily convince the reader of the truth of this statement.

As we shall explain in Appendix A, there can be no decidable system of axioms for the structure

$$\mathfrak{N} = \langle \mathbb{N};\ +^{\mathbb{N}},\ \cdot^{\mathbb{N}};\ 0^{\mathbb{N}},\ 1^{\mathbb{N}}\rangle.$$

This means, in other words, that every "decidable" system $\Sigma \subseteq \mathrm{Sent}(L)$ of sentences that hold in $\mathfrak{N}$ must necessarily be incomplete. One such system is the following:

$$\begin{array}{lll}
\mathrm{P}_1: & \forall x & x+1 \neq 0\\
\mathrm{P}_2: & \forall x,y & (x+1 \doteq y+1 \ \rightarrow\ x \doteq y)\\
\mathrm{P}_3: & \forall x & x+0 \doteq x\\
\mathrm{P}_4: & \forall x,y & x+(y+1) \doteq (x+y)+1\\
\mathrm{P}_5: & \forall x & x\cdot 0 \doteq 0\\
\mathrm{P}_6: & \forall x,y & x\cdot(y+1) \doteq x\cdot y+x\\
\mathrm{P}_\varphi: & \multicolumn{2}{l}{\varphi(v_0/0) \wedge \forall v_0\,(\varphi(v_0/v_0) \rightarrow \varphi(v_0/v_0+1)) \ \rightarrow\ \forall v_0\,\varphi(v_0/v_0).}
\end{array}$$

The axiom system

$$\Sigma_{\text{PA}} = \{P_1, \ldots, P_6\} \cup \{\forall P_\varphi \mid \varphi \in \text{Fml}(L)\}$$

is usually called the system of *Peano axioms* of arithmetic. The fact that $\forall P_\varphi$ holds in $\mathfrak{N}$ simply means that for every evaluation $h$ in $\mathfrak{N}$, the set

$$\{a \in \mathbb{N} \mid \mathfrak{N} \models \varphi\,[h(\tbinom{v_0}{a})]\}$$

has the following property: if 0 belongs to this set, and if whenever $n$ belongs to it, so does $n+1$, then every natural number belongs to it.

*9. Set theory*

Finally, we wish to present an axiom system for set theory – in fact, for *Zermelo–Fraenkel set theory*. The underlying language $L$ has (as in our first axiom system above, namely, in the case of orderings) only one, single relation symbol, of arity 2, which we write here as $\varepsilon$.[6] The following axioms are oriented toward the membership relation between sets. In order to prevent the axioms from becoming too unreadable, we shall at times introduce common abbreviations:

$\text{ZF}_1$: Axiom of extensionality
$\forall x, y\ (\forall z (z\, \varepsilon\, x \leftrightarrow z\, \varepsilon\, y) \rightarrow x \doteq y)$

If we write $x \subseteq y$ for $\forall z (z\, \varepsilon\, x \rightarrow z\, \varepsilon\, y)$, then $\text{ZF}_1$ is clearly equivalent to

$$\forall x, y\ (x \subseteq y \wedge y \subseteq x \rightarrow x \doteq y).$$

$\text{ZF}_1$ says that two sets possessing the same elements are equal.

$\text{ZF}_2$: Null set axiom
$\exists x\, \forall z\ \neg z\, \varepsilon\, x.$

This axiom says that a set possessing no elements exists.

$\text{ZF}_3$: Pair set axiom
$\forall u, v\, \exists x\, \forall z\ (z\, \varepsilon\, x \leftrightarrow (z \doteq u \vee z \doteq v))$

$\text{ZF}_4$: Union set axiom
$\forall u\, \exists x\, \forall z\ (z\, \varepsilon\, x \leftrightarrow \exists v (v\, \varepsilon\, u \wedge z\, \varepsilon\, v))$

---

[6] Note the distinction between the symbol $\varepsilon$, which belongs to the object language of set theory, and the symbol $\in$, which we reserve for the metalanguage.

$ZF_5$: Power set axiom
$\forall u\, \exists x\, \forall z\ (z \, \varepsilon\, x \leftrightarrow z \subseteq u)$

These last four axioms assert the existence of certain sets $x$. By $ZF_1$, each of these sets is uniquely determined, though in the case of $ZF_3$–$ZF_5$, $x$ depends on one or two other sets (or "parameters") $u$ and $v$. The $x$ in $ZF_2$, $ZF_3$, $ZF_4$ and $ZF_5$ is usually "denoted" by $\emptyset$ (the "empty set" or "null set"), $\{u,v\}$, $\bigcup u$ and $\mathrm{P}(u)$, respectively. This means that one adjoins to the language names (i.e. a constant symbol in the case of $ZF_2$, and function symbols in the cases of $ZF_3$–$ZF_5$) for these uniquely determined objects. This corresponds exactly to the usual progression in mathematics; it contributes to a faster and better understanding of formulae. But it is, in principle, not necessary. Instead of using the name of such a set, one could simply utilize a uniquely characterizing property of it, as follows.

Let $\psi$ be a formula in which $x$ occurs free, and for which we can prove from the axioms that $\psi$ holds for exactly one $x$ (though possibly depending on other variables occurring free in $\psi$, as mentioned above). For this uniquely determined set we introduce a new symbol, such as $\emptyset$ or $\{u,v\}$, as above. But we do not adjoin this new symbol to the language of set theory – the $\varepsilon$-language. Then, if $\varphi$ is an $\varepsilon$-formula with free variable $y$, we shall write $\varphi(y/\emptyset)$ to stand for the following $\varepsilon$-formula:

$$\exists x\, (\psi(x) \wedge \varphi(y/x)),$$

where $\psi(x)$ is the formula $\forall z\ \neg z \in x$. Here we assume that $x$ is free for $y$ in $\varphi$; otherwise, instead of $x$ we would utilize some variable that occurs in neither $\psi$ nor $\varphi$. In the case of $\{u,v\}$, $\psi$ would be $\forall z\, (z\, \varepsilon\, x \leftrightarrow (z \doteq u \vee z \doteq v))$. As the last example shows, $\psi$ can have, besides $x$, additional free variables, as mentioned above. In the following axioms we shall make use of these abbreviations.

From the previous axioms it follows that if $x$ and $y$ are sets, then so are $\{x,x\}$ and $\bigcup\{x,y\}$. In the case of $\{x,x\}$ we also write $\{x\}$, and in the case of $\bigcup\{x,y\}$ we also write $x \cup y$. Furthermore, it is customary to write $x'$ for $x \cup \{x\}$. Now the next axiom can be formulated using these notations:

$ZF_6$: Infinity axiom
$\exists x\, (\emptyset \, \varepsilon\, x \wedge \forall z (z \, \varepsilon\, x \rightarrow z' \, \varepsilon\, x)).$

This axiom ensures the existence of an infinite set. This set is, however, not uniquely determined.

The next axiom says that the image of a set under a function is again a set. Here the function is described (implicitly) by a formula $\varphi$, in which the variables $u$ and $v$ occur free. Here and elsewhere we shall use notation such as $\varphi(u,v)$ to emphasize the possibility that $u$ and $v$ occur free in $\varphi$; this notation does not exclude the possibility that other variables might also occur free in $\varphi$.

ZF$_7$: Replacement axiom

$$\forall y\,(\forall u\,(u\,\varepsilon\,y \to \exists^{=1}v\,\varphi(u,v)) \;\to\; \exists x \forall v\,(v\,\varepsilon\,x \leftrightarrow \exists u\,(u\,\varepsilon\,y \wedge \varphi(u,v)))).$$

Thus $x$ is the image of $y$ under the mapping described by $\varphi$. By ZF$_1$, this $x$ is uniquely determined. If $\psi$ is a formula in which the variable $v$ occurs free, then the abbreviation $\exists^{=1}v\,\psi(v)$, used in ZF$_7$ above, stands for

$$\exists v\,(\psi(v) \wedge \forall u\,(\psi(v/u) \to u \doteq v)),$$

where $u$ is a variable not occurring in $\psi$.

The last axiom of Zermelo–Fraenkel set theory is usually

ZF$_8$: Foundation axiom

$$\forall x\,(x \neq \emptyset \;\to\; \exists z\,(z\,\varepsilon\,x \,\wedge\, z \cap x \doteq \emptyset)).$$

Here the notation $z \cap x$ (for "intersection") denotes a set whose existence is proved using ZF$_7$, and whose uniqueness is guaranteed by ZF$_1$.

Usually the so-called Axiom of Choice is added to the above ZF-axioms; we present it in the following form.

ZF$_9$: Axiom of Choice

$$\forall y, u\,(u \subseteq \mathrm{P}(y) \wedge \forall z_1, z_2\,(z_1\,\varepsilon\,u \wedge z_2\,\varepsilon\,u \to z_1 \neq \emptyset \wedge (z_1 \doteq z_2 \vee z_1 \cap z_2 \doteq \emptyset))$$
$$\to \exists x\,(x \subseteq y \wedge \forall z\,(z\,\varepsilon\,u \to \exists v\, x \cap z \doteq \{v\}))).$$

This axiom says that to every system $u$ of nonempty, pairwise disjoint subsets of a given set $y$, there exists a "system $x$ of representatives", i.e. $x$ contains, for every element $z$ of $u$, exactly one element $v\,\varepsilon\,z$, the representative for $z$.[7]

By consulting the appropriate literature (e.g. [Levy, 1979–2002]), the reader should be able to convince himself that the entirety of mathematics may actually be built up in Zermelo–Fraenkel set theory.

## 1.7 Exercises for Chapter 1

**1.7.1.** Notation: $F(x)$ $x$ is female

$C(x,y)$ $x$ is a child of $y$

$M(x,y)$ $x$ and $y$ are married.

Formalize the following statements:

(a) $x$ is single.

(b) $x$ is the brother-in-law of $y$.

[7] The system $x$ may also contain other elements of $y$, not belonging to any $z\,\varepsilon\,u$.

(c) No two of the three grandchildren of $x$ are siblings.
Here, the variables $x, y, \ldots$ range over all people.

**1.7.2.** Notation:

| | |
|---|---|
| $P(x)$ | $x$ is a point |
| $G(x)$ | $x$ is a straight line |
| $E(x)$ | $x$ is a plane |
| $I(x,y)$ | $x$ lies on $y$ |

Formalize the following statements:

(a) For every two points, there is a straight line on which these two points lie.

(b) Through two distinct points there is at most one straight line.

(c) For every three points that do not lie on a common line, there is exactly one plane on which these three points lie.

Here, the variables $x, y, \ldots$ range over all points, lines and planes simultaneously.

**1.7.3.** Notation:

| | |
|---|---|
| $R(x,y)$ | $x$ is less than or equal to $y$ |
| $b(x)$ | absolute value of $x$ |
| $p(x,y)$ | $x$ times $y$ |
| $s(x,y)$ | $x$ plus $y$ |
| $d(x,y)$ | $x$ minus $y$ |
| $f(x)$ | one-place function of $x$ |
| $e$ | one |
| $n$ | zero |

Formalize the following statements:

(a) $f$ is differentiable at 1.

(b) $f$ is uniformly continuous on the interval $[0,1]$.

Here, the variables $x, y, \ldots$ vary over $\mathbb{R}$.

**1.7.4.** Let $L$ be a formal language with the one-place function symbol $|\cdot|$, the two-place function symbols $f, +, \cdot, -$, the two-place relation symbol $<$ and the constant symbols $0, 1, x_0, y_0$.

Formalize in $L$ the Implicit Function Theorem: If the function $f$ vanishes at the point $(x_0, y_0)$, but the partial derivative of $f$ with respect to the second variable does not, then there are open neighbourhoods $U$ of $x_0$ and $V$ of $y_0$ such that, for each $x \in U$, there is exactly one $y \in V$ with $f(x,y) = 0$.

**1.7.5.** Which of the following implications are tautologies?

(a) $(\alpha \to (\beta \to \gamma)) \to (\alpha \wedge \beta \to \gamma)$

(b) $(\alpha \to (\beta \to \gamma)) \to (\alpha \vee \beta \to \gamma)$

(c) $(\alpha \to (\beta \to \gamma)) \to ((\alpha \to \beta) \to \gamma)$

(d) $(\alpha \wedge (\beta \to \gamma)) \to (\alpha \wedge \beta \to \gamma)$

(e) $(\alpha \to \gamma) \wedge (\beta \to \gamma) \to (\alpha \vee \beta \to \gamma)$

**1.7.6.** Let $L$ be a formal language with the one-place function symbol $-$, the two-place function symbols $+$ and $\cdot$, and the constant symbols 0 and 1. Let $\Sigma$ be the set of the following four axioms:

$$\begin{aligned}
&\forall x,y,z\ (x+y)+z \doteq x+(y+z)\\
&\forall x,y,z\ (x+y)\cdot z \doteq x\cdot z+y\cdot z\\
&\forall x \quad x+0 \doteq x\\
&\forall x \quad x+(-x) \doteq 0
\end{aligned}$$

Show that $\Sigma \vdash (0\cdot 0 \doteq 0)$.

**1.7.7.** Show:
(i) $\emptyset \vdash \forall x(\varphi \to \psi) \to (\forall x\varphi \to \forall x\psi)$.
*Hint*: It could be helpful to apply the tautologies

$$(\alpha \to (\varphi \to \psi)) \to ((\beta \to \varphi) \to (\alpha \wedge \beta \to \psi)) \quad \text{and}$$
$$(\alpha \wedge \beta \to \psi) \to (\alpha \to (\beta \to \psi))$$

with suitable formulae $\alpha$ and $\beta$.
(ii) $\emptyset \vdash \forall x(\varphi \to \psi) \to (\exists x\varphi \to \psi)$ in case $x \notin \mathrm{Fr}(\psi)$.

**1.7.8.** Show that for every quantifier-free formula $\varphi$ there is a likewise quantifier-free formula $\psi$, with the same free variables, in conjunctive (respectively, disjunctive) normal form, for which $\emptyset \vdash (\varphi \leftrightarrow \psi)$ holds.

**1.7.9.** Show that

$$\Sigma \vdash (\varphi \to \psi)$$

does not follow from $\Sigma \cup \{\varphi\} \vdash \psi$.
*Hint*: In order to show that a formula is not provable from an axiom system, one can give a mathematical domain (such as the real numbers) in which all the axioms hold, but not the formula in question. Furthermore, by the Deduction Theorem, $\varphi$ must contain a free variable, so that one can construct such a counterexample.

**1.7.10.** For formulae $\varphi$ and $\psi$, prove:

(a) $\emptyset \vdash (\exists x\,\varphi \vee \exists x\,\psi) \leftrightarrow \exists x\,(\varphi \vee \psi)$
(b) $\emptyset \vdash (\forall x\,\varphi \vee \forall y\,\psi) \to \forall x\,(\varphi \vee \psi)$
(c) $\emptyset \nvdash \forall x\,(\varphi \vee \psi) \to (\forall x\,\varphi \vee \forall x\,\psi)$

**1.7.11.** Show that for every formula $\varphi$ there is a formula $\psi$, in prenex normal form, for which $\mathrm{Fr}(\varphi) = \mathrm{Fr}(\psi)$ and $\emptyset \vdash (\varphi \leftrightarrow \psi)$.

**1.7.12.** Give a formula $\varphi$ in the language $(+,\cdot,0,1)$ of field theory, with $\mathrm{Fr}(\varphi) = \{x_0,\dots,x_n\}$, such that an $(n+1)$-tuple $(a_0,\dots,a_n)$ of elements of an arbitrary infinite field satisfies the formula if and only if the polynomial $a_0 + a_1X + \dots + a_nX^n$ is irreducible over that field.

**1.7.13.** Let $L = (+,\cdot,0,1)$ be the language of field theory.
For which fields $\mathscr{A}$ is the universe $|\mathscr{A}|$ equal to the set

$$\{ t^{\mathscr{A}} \mid t \text{ is a constant term} \}?$$

**1.7.14.** Let $L$ be a formal language. We say that a class $\mathscr{M}$ of $L$-structures is (*finitely*) *axiomatized* by a (finite) set $\Sigma$ of $L$-sentences if $\mathscr{M}$ consists exactly of the models of $\Sigma$, i.e. $\mathscr{M}$ is the model class of $\Sigma$ as defined in Section 3.1.

(i) Show that the class $\mathscr{M}$ of all finite $L$-structures cannot be axiomatized with $L$-sentences.

*Hint*: Suppose that there were an axiomatization. Add suitable additional sentences to these axioms, and apply the Finiteness Theorem to obtain a contradiction.

(ii) Show that the class of infinite $L$-structures can be axiomatized with infinitely many $L$-sentences, but not, however, with only finitely many.

**1.7.15.** Let the language $L$ consist only of the two-place relation symbol $>$. We call an $L$-structure $\mathfrak{A}$ a *well-ordered set* if $>^{\mathfrak{A}}$ is a strict linear ordering on $|\mathfrak{A}|$, with respect to which there is no infinite descending sequence

$$a_0 >^{\mathfrak{A}} a_1 >^{\mathfrak{A}} \cdots >^{\mathfrak{A}} a_n >^{\mathfrak{A}} \cdots .$$

Show that the class $\mathscr{M}$ of well-ordered sets cannot be axiomatized[9] by a set of $L$-sentences.

*Hint*: Extend the language by adding constants, if necessary, and then proceed similarly as in Exercise 1.7.14(i).

**1.7.16.** Let $E$ be a two-place relation symbol, and $L = (E)$ the *language of graphs*. By a *graph* we mean a (possibly infinite) $L$-structure $\mathfrak{A}$, in which the two-place relation $E^{\mathfrak{A}}$ is symmetric and irreflexive. (One imagines the universe as a set of points in which any two points $x,y$ are connected by a line whenever $E^{\mathfrak{A}}(x,y)$ holds.)

Which of the following classes of $L$-structures are axiomatizable?[9] Either give an axiom system, or show that there can be none.

(i) The class of all graphs.
(ii) The class of all $L$-structures that are not graphs.
(iii) The class of all finite graphs (i.e. graphs with a finite universe).
(iv) The class of all infinite graphs.
(v) The class of all connected graphs.

Here, a graph is called *connected* if for every two points $x$ and $y$, there is a finite path from $x$ to $y$, i.e. a finite sequence of edges that begins at $x$ and ends at $y$.

**1.7.17.** Let $L = (+,\cdot,0,1)$ be the language of rings. Add a three-place relation symbol $M$. For an arbitrary ring $\mathfrak{A}$ and $a,b,c \in |\mathfrak{A}|$, we now define $M^{\mathfrak{A}}(a,b,c) :\Leftrightarrow a \cdot^{\mathfrak{A}} b = c$.

Give a formula without function symbols that expresses precisely the commutativity of the ring multiplication.

---

[9] Recall the definition of "axiomatized" given in Exercise 1.7.14.

**1.7.18.** Let $L$ be a formal language. For each function symbol $f_i$ of $L$, we add a new relation symbol $F_i$ with arity increased by one. In this way we obtain the extended language $L'$. We transform an $L$-structure $\mathfrak{A}$ into an $L'$-structure $\mathfrak{A}'$ by the following definition:

$$F_i^{\mathfrak{A}'}(a_1,\ldots,a_s,a) :\Leftrightarrow f_i^{\mathfrak{A}}(a_1,\ldots,a_s) = a,$$

for elements $a_1,\ldots,a_s,a \in |\mathfrak{A}|$.

Show that one can assign to each $L$-formula $\varphi$ an $L'$-formula $\varphi'$ with the following properties:

(i) No function symbols occur in $\varphi'$.

(ii) For arbitrary $L$-structures $\mathfrak{A}$ and arbitrary evaluations $h$ of the variables, we have:

$$\mathfrak{A} \models \varphi[h] \;\Leftrightarrow\; \mathfrak{A}' \models \varphi'[h].$$

*Hint*: First consider prime formulae $\varphi$ of the form $t \doteq x$, with $t$ a term and $x$ a variable, and define $\varphi'$ recursively according to the construction of $t$. Then we define the assignment $\varphi \mapsto \varphi'$ recursively according the construction of formulae.

# Chapter 2
# Model Constructions

In this chapter we shall introduce various methods for the construction of models of an axiom system $\Sigma$ of sentences in a formal language $L$. In Chapter 1 we have already encountered a method, in the form of the so-called term-models, for obtaining at least one model of $\Sigma$. Having presented that "absolute" construction, we shall now present a series of "relative" constructions. These relative methods allow us to start with one or more given models of $\Sigma$, and to produce a new model. The methods considered here do not (as often occurs in mathematics) depend on the particular axiom system $\Sigma$ (e.g. the direct product of groups is again a group, while the analogue of this for fields does not hold); rather, our methods will work in every case. This will be guaranteed by the fact that an $L$-structure $\mathfrak{A}'$ obtained by these methods from an $L$-structure $\mathfrak{A}$ is *elementarily equivalent* to $\mathfrak{A}$; i.e.

*every L-sentence $\varphi$ that holds in $\mathfrak{A}$ also holds in $\mathfrak{A}'$, and conversely.* (2.0.0.1)

Therefore, if $\mathfrak{A}$ is a model of $\Sigma$, then so is $\mathfrak{A}'$, independent of which axiom system $\Sigma \subseteq \mathrm{Sent}(L)$ we are working with.

Elementary equivalence between two $L$-structures $\mathfrak{A}$ and $\mathfrak{A}'$ (written $\mathfrak{A} \equiv \mathfrak{A}'$) is obviously an equivalence relation on the class of all $L$-structures; i.e. for $L$-structures $\mathfrak{A}, \mathfrak{A}', \mathfrak{A}''$ we have:

$$\begin{aligned}&\mathfrak{A} \equiv \mathfrak{A}\\ &\mathfrak{A} \equiv \mathfrak{A}' \quad \text{implies} \quad \mathfrak{A}' \equiv \mathfrak{A}\\ &\mathfrak{A} \equiv \mathfrak{A}' \text{ and } \mathfrak{A}' \equiv \mathfrak{A}'' \quad \text{imply} \quad \mathfrak{A} \equiv \mathfrak{A}''.\end{aligned}$$

For $L$-structures $\mathfrak{A}$ and $\mathfrak{A}'$, we shall introduce the notion of an $L$-isomorphism, $\mathfrak{A} \cong \mathfrak{A}'$, generalizing the usual concept of isomorphism in algebra. We shall see that isomorphism always entails elementary equivalence, i.e.

$$\mathfrak{A} \cong \mathfrak{A}' \quad \text{implies} \quad \mathfrak{A} \equiv \mathfrak{A}',$$

A. Prestel, C.N. Delzell, *Mathematical Logic and Model Theory*, Universitext,
DOI 10.1007/978-1-4471-2176-3_3, © Springer-Verlag London Limited 2011

while the converse is false in general. For "finite" $L$-structures (i.e. those whose universe is finite), the converse always holds, but not for infinite ones. We shall see that the $\equiv$-equivalence class of an infinite $L$-structure contains $L$-structures of arbitrarily large cardinality. (Observe that isomorphism, by contrast, naturally implies equality of cardinality of the universes.)

The set-theoretic concepts used in this part of the book can be found in any book on set theory (e.g. [Levy, 1979–2002]). We denote the cardinal number of a set $A$ by $\mathrm{card}(A)$.

## 2.1 Term Models

Let $L = (\lambda, \mu, K)$ be a formal language, as introduced in Section 1.2. We call the cardinal number

$$\kappa_L := \max\{\aleph_0,\ \mathrm{card}(I),\ \mathrm{card}(J),\ \mathrm{card}(K)\} \tag{2.1.0.1}$$

the *cardinality of the language L*. If $\mathfrak{A}$ is an $L$-structure, then the *cardinality of* $\mathfrak{A}$ will be understood to be the cardinality of the universe $|\mathfrak{A}|$ of $\mathfrak{A}$.

Our goal in this section is to prove the following theorem:

**Theorem 2.1.1.** *If the set* $\Sigma \subseteq \mathrm{Sent}(L)$ *of sentences has a model of infinite cardinality, then for every cardinal number* $\kappa \geq \kappa_L$*,* $\Sigma$ *has a model of cardinality* $\kappa$*.*

Before we prove this theorem, we wish to make some preliminary observations.

For a language $L = (\lambda, \mu, K)$ with arity-functions $\lambda : I \to \mathbb{N}$ and $\mu : J \to \mathbb{N}$, we have

$$\mathrm{card}(\mathrm{Tm}(L)) = \max\{\aleph_0,\ \mathrm{card}(J),\ \mathrm{card}(K)\} \tag{2.1.1.1}$$
$$\mathrm{card}(\mathrm{Sent}(L)) = \mathrm{card}(\mathrm{Fml}(L)) = \kappa_L. \tag{2.1.1.2}$$

These follow from the well-known fact that the set of all finite sequences with values in an infinite set $M$ has the same cardinality as $M$. For (2.1.1.1) we apply this fact to the set

$$M := \mathrm{Vbl} \cup \{f_j \mid j \in J\} \cup \{c_k \mid k \in K\} \cup \{,\} \cup \{)\} \cup \{(\},$$

whose cardinality is just $\max\{\aleph_0,\ \mathrm{card}(J),\ \mathrm{card}(K)\}$. Now (2.1.1.1) follows, since, on the one hand, every term is a finite sequence from $M$ (proving $\leq$ in (2.1.1.1)), and, on the other hand, every $v_n$, every $c_k$, and every $f_j(v_0, \ldots, v_0)$ (for $j \in J$) is a term (proving $\geq$).

For (2.1.1.2) we apply the above-mentioned fact to the set

$$M := \mathrm{Tm}(L) \cup \{R_i \mid i \in I\} \cup \{\neg\} \cup \{\wedge\} \cup \{\forall\} \cup \{\doteq\} \cup \{,\} \cup \{)\} \cup \{(\},$$

whose cardinality is just $\kappa_L$, by (2.1.1.1) and (2.1.0.1). On the one hand, $\mathrm{Sent}(L) \subseteq \mathrm{Fml}(L)$, and every formula is a finite sequence of elements of $M$. From this we get

$$\mathrm{card}(\mathrm{Sent}(L)) \leq \mathrm{card}(\mathrm{Fml}(L)) \leq \kappa_L.$$

On the other hand, $\forall t \doteq t$ and $\forall v_0 R_i(v_0, \ldots, v_0)$ (for $i \in I$) are sentences in $L$, from which we get

$$\kappa_L \leq \mathrm{card}(\mathrm{Sent}(L)). \qquad \square$$

Using these observations we now wish to give an upper bound on the cardinality of the term model constructed (in Section 1.4) from a set $\Sigma \subseteq \mathrm{Sent}(L)$ of sentences. Thus, we must bound the cardinality of the set

$$\mathrm{CT}/\approx$$

of equivalence classes of constant terms in the extended language $L'$ (1.4.5.4). Since

$$\mathrm{card}(\mathrm{CT}/\approx) \leq \mathrm{card}(\mathrm{Tm}(L')) \leq \kappa_{L'},$$

by (2.1.1.1), it suffices to calculate $\kappa_L$. Recall (1.4.3.2) that the language $L'$ was constructed as the union of an ascending chain

$$L = L_0 \subseteq L_1 \subseteq \cdots \subseteq L_{n-1} \subseteq L_n \subseteq \cdots$$

of languages. If we can succeed in showing that

$$\kappa_{L_n} \leq \kappa_L \qquad (2.1.1.3)$$

for every $n \in \mathbb{N}$, then we will obviously have $\kappa_{L'} \leq \kappa_L$ and thus $\kappa_{L'} = \kappa_L$. We obtain (2.1.1.3) by induction on $n$. For $n = 0$ it is trivial. If it holds for $n - 1$, then it also holds for $n$: the language $L_n$ is constructed from $L_{n-1}$ by expanding the index set $K_{n-1}$ by a set $M_n$, which is equipotent to a subset of $\mathrm{Sent}(L_{n-1})$ (recall (1.4.3.1)). Therefore

$$\kappa_{L_n} \leq \max\{\kappa_{L_{n-1}}, \mathrm{card}(M_n)\} \leq \kappa_{L_{n-1}} \leq \kappa_L,$$

using (2.1.1.2). We have proved:

$$\mathrm{card}(\mathrm{CT}/\approx) \leq \kappa_L. \qquad \square \qquad (2.1.1.4)$$

We are now in a position to give the proof of Theorem 2.1.1.

*Proof* (of Theorem 2.1.1): Starting with the given language $L$ and any cardinal number $\kappa \geq \kappa_L$, we define an extension language $L_\kappa$ by

$$I_\kappa = I, \quad J_\kappa = J, \quad K_\kappa = K \cup \{\nu \mid \nu < \kappa\},$$

where we assume the last union is disjoint. An $L_\kappa$-structure is then an $L$-structure together with the interpretations of the new constant symbols $c_\nu$ for $\nu < \kappa$.

In the language $L_\kappa$ we consider the set of sentences

$$\Sigma_\kappa := \Sigma \cup \{c_\nu \neq c_\mu \mid \nu < \mu < \kappa\}.$$

This set is consistent by the Finiteness Theorem 1.5.6: every finite subset $\Pi$ of $\Sigma_\kappa$ is contained in a set of the form

$$\Sigma \cup \{c_{\nu_1} \neq c_{\mu_1}, \ldots, c_{\nu_n} \neq c_{\mu_n}\},$$

for various ordinal numbers $\nu_i < \mu_i < \kappa$, $i = 1, 2, \ldots, n$. This set has a model, since by hypothesis $\Sigma$ has an infinite model $\mathfrak{A}$. Indeed, it suffices to adjoin to $\mathfrak{A}$ an interpretation of the new constant symbols $c_\nu$, for $\nu < \kappa$, in such a way that for $i = 1, \ldots, n$, the two constant symbols $c_{\nu_i}$ and $c_{\mu_i}$ are interpreted differently. This is possible, since the universe $|\mathfrak{A}|$ is infinite.

Now we let $\mathfrak{A}'$ be the term model constructed from $\Sigma_\kappa$ and $L_\kappa$ (rather than from $\Sigma$ and $L$, respectively) in (1.5.2.2) (actually, the main steps of the construction were presented already in Section 1.4). Then $\mathfrak{A}'$ is an $L$-structure together with interpretations of the constant symbols $c_\nu$ (for $\nu < \kappa$) as well as of the new constant symbols arising in the passage from $L_\kappa$ to $L'$ in (the proof of) Theorem 1.4.2. The restriction $\mathfrak{B}$ of $\mathfrak{A}'$ to the language $L_\kappa$, which is obtained simply by dropping the interpretations of the additional constant symbols, is then clearly a model of $\Sigma$. It remains to prove that $|\mathfrak{B}|$ has cardinality $\kappa$.

Since $\mathfrak{A}'$ is a model of $\{c_\nu \neq c_\mu \mid \nu < \mu < \kappa\}$, the interpretation

$$c_\nu \mapsto c_\nu^{\mathfrak{A}'}$$

induces an injection from $\{c_\nu \mid \nu < \kappa\}$, and thus from $\{\nu \mid \nu < \kappa\}$, into the universe of $\mathfrak{A}'$, which is at the same time the universe of $\mathfrak{B}$..Therefore

$$\kappa \leq \operatorname{card}(|\mathfrak{B}|). \tag{2.1.1.5}$$

On the other hand,

$$\begin{aligned} \operatorname{card}(|\mathfrak{B}|) &\leq \kappa_{L_\kappa} && \text{by (2.1.1.4)} \\ &= \max\{\aleph_0, \operatorname{card}(I), \operatorname{card}(J), \operatorname{card}(K \cup \{\nu \mid \nu < \kappa\}\} \\ &\leq \max\{\kappa_L, \kappa\} = \kappa. && (2.1.1.6) \end{aligned}$$

From (2.1.1.5) and (2.1.1.6) we obtain $\operatorname{card}(|\mathfrak{B}|) = \kappa$. □

From this theorem we obtain easily that the $\equiv$-equivalence class of any infinite $L$-structure $\mathfrak{A}$ contains $L$-structures of arbitrarily large cardinality. To prove this, simply consider the following set of $L$-sentences:

$$\operatorname{Th}(\mathfrak{A}) = \{\varphi \in \operatorname{Sent}(L) \mid \mathfrak{A} \models \varphi\}. \tag{2.1.1.7}$$

$\operatorname{Th}(\mathfrak{A})$ was in (1.6.0.1) called *the L-theory of* $\mathfrak{A}$. Every model $\mathfrak{B}$ of $\operatorname{Th}(\mathfrak{A})$ is obviously elementarily equivalent to $\mathfrak{A}$ (2.0.0.1). Indeed, from $\mathfrak{A} \models \varphi$ follows $\varphi \in \operatorname{Th}(\mathfrak{A})$ and hence $\mathfrak{B} \models \varphi$. Conversely, from $\mathfrak{A} \not\models \varphi$ follows naturally $\neg\varphi \in \operatorname{Th}(\mathfrak{A})$ and hence $\mathfrak{B} \not\models \varphi$. Now applying Theorem 2.1.1 to the set $\Sigma = \operatorname{Th}(\mathfrak{A})$, we get

**Corollary 2.1.2.** *For every L-structure $\mathfrak{A}$ of infinite cardinality, and for every cardinal number $\kappa \geq \kappa_L$, there is an L-structure $\mathfrak{B}$ of cardinality $\kappa$ that is elementarily equivalent to $\mathfrak{A}$.* □

As we show in the next section, this has the consequence that no infinite $L$-structure can be characterized, up to isomorphism, by means of an axiom system $\Sigma \subseteq \mathrm{Sent}(L)$.

## 2.2 Morphisms of Structures

Given two $L$-structures $\mathfrak{A}$ and $\mathfrak{A}'$, we call a mapping $\tau : |\mathfrak{A}| \to |\mathfrak{A}'|$ an *L-isomorphism* (or simply an *isomorphism*, if the language is clear) if the following conditions are satisfied:

($I_1$) $\tau : |\mathfrak{A}| \to |\mathfrak{A}'|$ is bijective;

($I_2$) $R_i^{\mathfrak{A}}(a_1,\ldots,a_{\lambda(i)}) \Leftrightarrow R_i^{\mathfrak{A}'}(\tau(a_1),\ldots,\tau(a_{\lambda(i)}))$
for all $i \in I$ and for all $a_1,\ldots,a_{\lambda(i)} \in |\mathfrak{A}|$;

($I_3$) $\tau\big(f_j^{\mathfrak{A}}(a_1,\ldots,a_{\mu(j)})\big) = f_j^{\mathfrak{A}'}(\tau(a_1),\ldots,\tau(a_{\mu(j)}))$
for all $j \in J$ and for all $a_1,\ldots,a_{\mu(j)} \in |\mathfrak{A}|$;

($I_4$) $\tau\big(c_k^{\mathfrak{A}}\big) = c_k^{\mathfrak{A}'}$ for all $k \in K$.

If all of these conditions are fulfilled, then we also write

$$\tau : \mathfrak{A} \leftrightarrow \mathfrak{A}',$$

for short. If $\tau : |\mathfrak{A}| \to |\mathfrak{A}'|$ fulfils conditions ($I_2$), ($I_3$) and ($I_4$), and is injective (but not necessarily surjective), then we speak of an *(L-)monomorphism* or an *(L-)embedding*, and we write

$$\tau : \mathfrak{A} \hookrightarrow \mathfrak{A}'.$$

If there is an isomorphism between $\mathfrak{A}$ and $\mathfrak{A}'$, then we call $\mathfrak{A}$ and $\mathfrak{A}'$ *(L-)isomorphic*, and we write

$$\mathfrak{A} \cong \mathfrak{A}'.$$

If there is an embedding from $\mathfrak{A}$ into $\mathfrak{A}'$, then we call $\mathfrak{A}$ *(L-)embeddable* in $\mathfrak{A}'$, and we write

$$\mathfrak{A} \hookrightarrow \mathfrak{A}'.$$

An isomorphism from $\mathfrak{A}$ to itself is usually called an *(L-)automorphism*.

For later reference we remark that we can bring conditions ($I_3$) and ($I_4$) into a form analogous to that of ($I_2$), namely:

$$(\mathrm{I}_3') \qquad f_j^{\mathfrak{A}}(a_1,\ldots,a_{\mu(j)}) = d \quad \text{implies}$$
$$f_j^{\mathfrak{A}'}(\tau(a_1),\ldots,\tau(a_{\mu(j)})) = \tau(d), \quad \text{for all } j \in J \text{ and } a_1,\ldots,a_{\mu(j)}, d \in |\mathfrak{A}|;$$

$$(\mathrm{I}_4') \qquad c_k^{\mathfrak{A}} = d \quad \text{implies}$$
$$c_k^{\mathfrak{A}'} = \tau(d) \quad \text{for all } k \in K \text{ and } d \in |\mathfrak{A}|.$$

The equivalence with ($\mathrm{I}_3$) and ($\mathrm{I}_4$) is immediately clear. The reason that an implication suffices here (in contrast to the equivalence needed in ($\mathrm{I}_2$)) is that the $d$ in ($\mathrm{I}_3'$), for example, is already uniquely determined by the arguments $a_1,\ldots,a_{\mu(j)}$.

**Theorem 2.2.1.** *Let $\mathfrak{A}$ and $\mathfrak{A}'$ be L-structures, and $\tau : \mathfrak{A} \leftrightarrow \mathfrak{A}'$ an isomorphism. Then for every L-formula $\varphi$ and every evaluation $h$ in $\mathfrak{A}$,*

$$\mathfrak{A} \models \varphi[h] \quad \textit{iff} \quad \mathfrak{A}' \models \varphi[\tau \circ h]. \tag{2.2.1.1}$$

*In particular, $\mathfrak{A}$ and $\mathfrak{A}'$ are elementarily equivalent.*

*Proof*: Since $h$ is an evaluation in $\mathfrak{A}$, $\tau \circ h$ is an evaluation in $\mathfrak{A}'$. Set $h' = \tau \circ h$. We begin by proving, for terms $t \in \mathrm{Tm}(L)$,

$$\tau(t^{\mathfrak{A}}[h]) = t^{\mathfrak{A}'}[h'], \tag{2.2.1.2}$$

by induction on the construction of $t$: When $t$ is a constant symbol $c_k$, (2.2.1.2) is just the condition ($\mathrm{I}_4$). When $t$ is a variable $x$, (2.2.1.2) is trivial, in view of

$$\tau(x^{\mathfrak{A}}[h]) = \tau(h(x)) = h'(x) = x^{\mathfrak{A}'}[h'].$$

When $t$ is $f_j(t_1,\ldots,t_{\mu(j)})$ we prove (2.2.1.2) as follows:

$$\begin{aligned}
\tau(f_j(t_1,\ldots,t_{\mu(j)}))^{\mathfrak{A}}[h] &= \tau(f_j^{\mathfrak{A}}(t_1^{\mathfrak{A}}[h],\ldots,t_{\mu(j)}^{\mathfrak{A}}[h])) && \text{by (1.5.0.4)}\\
&= f_j^{\mathfrak{A}'}(\tau(t_1^{\mathfrak{A}}[h]),\ldots,\tau(t_{\mu(j)}^{\mathfrak{A}}[h])) && \text{by } (\mathrm{I}_3)\\
&= f_j^{\mathfrak{A}'}(t_1^{\mathfrak{A}'}[h'],\ldots,t_{\mu(j)}^{\mathfrak{A}'}[h']) && \text{by ind. hyp. (2.2.1.2)}\\
&= f_j(t_1,\ldots,t_{\mu(j)})^{\mathfrak{A}'}[h'] && \text{by (1.5.0.4).}
\end{aligned}$$

We can now prove (2.2.1.1) by induction on the construction of the formula $\varphi$. For the atomic formula $t_1 \doteq t_2$ we have:

$$\begin{aligned}
\mathfrak{A} \models t_1 \doteq t_2[h] \quad &\text{iff} \quad t_1^{\mathfrak{A}}[h] = t_2^{\mathfrak{A}}[h] && \text{by (1.5.0.5)}\\
&\text{iff} \quad \tau(t_1^{\mathfrak{A}}[h]) = \tau(t_2^{\mathfrak{A}}[h]) && \tau \text{ injective } (\mathrm{I}_1)\\
&\text{iff} \quad t_1^{\mathfrak{A}'}[h'] = t_2^{\mathfrak{A}'}[h'] && \text{by (2.2.1.2)}\\
&\text{iff} \quad \mathfrak{A}' \models t_1 \doteq t_2[h'] && \text{by (1.5.0.5).}
\end{aligned}$$

For the atomic formula $R_i(t_1,\ldots,t_{\lambda(i)})$ we have:

$$
\begin{aligned}
\mathfrak{A} \models R_i(t_1,\ldots,t_{\lambda(i)})\,[h] &\text{ iff } R_i^{\mathfrak{A}}\big(t_1^{\mathfrak{A}}[h],\ldots,t_{\lambda(i)}^{\mathfrak{A}}[h]\big) && \text{by (1.5.0.6)}\\
&\text{ iff } R_i^{\mathfrak{A}'}\big(\tau(t_1^{\mathfrak{A}}[h]),\ldots,\tau(t_{\lambda(i)}^{\mathfrak{A}}[h])\big) && \text{by } (\mathrm{I}_2)\\
&\text{ iff } R_i^{\mathfrak{A}'}\big(t_1^{\mathfrak{A}'}[h'],\ldots,t_{\lambda(i)}^{\mathfrak{A}'}[h']\big) && \text{by (2.2.1.2)}\\
&\text{ iff } \mathfrak{A}' \models R_i(t_1,\ldots,t_{\lambda(i)})\,[h'] && \text{by (1.5.0.6).}
\end{aligned}
$$

For the formula $\neg\psi$ we have:

$$
\begin{aligned}
\mathfrak{A} \models \neg\psi\,[h] &\text{ iff } \mathfrak{A} \not\models \psi\,[h] && \text{by (1.5.0.7)}\\
&\text{ iff } \mathfrak{A}' \not\models \psi\,[h'] && \text{by ind. hyp. (2.2.1.1)}\\
&\text{ iff } \mathfrak{A}' \models \neg\psi\,[h'] && \text{by (1.5.0.7).}
\end{aligned}
$$

For the formula $(\psi_1 \wedge \psi_2)$ we have, analogously:

$$
\begin{aligned}
\mathfrak{A} \models (\psi_1 \wedge \psi_2)\,[h] &\text{ iff } (\mathfrak{A} \models \psi_1\,[h] \text{ and } \mathfrak{A} \models \psi_2\,[h]) && \text{by (1.5.0.8)}\\
&\text{ iff } (\mathfrak{A}' \models \psi_1\,[h'] \text{ and } \mathfrak{A}' \models \psi_2\,[h']) && \text{by ind. hyp. (2.2.1.1)}\\
&\text{ iff } \mathfrak{A}' \models (\psi_1 \wedge \psi_2)\,[h'] && \text{by (1.5.0.8).}
\end{aligned}
$$

Finally, in the case of a formula $\forall x\,\psi$ we have:

$$
\begin{aligned}
\mathfrak{A} \models \forall x\,\psi\,[h] &\text{ iff } \mathfrak{A} \models \psi\left[h\binom{x}{a}\right] && \text{for all } a \in |\mathfrak{A}| && \text{by (1.5.0.9)}\\
&\text{ iff } \mathfrak{A}' \models \psi\left[\tau\left(h\binom{x}{a}\right)\right] && \text{for all } a \in |\mathfrak{A}| && \text{by ind. hyp. (2.2.1.1)}\\
&\text{ iff } \mathfrak{A}' \models \psi\left[h'\binom{x}{\tau(a)}\right] && \text{for all } a \in |\mathfrak{A}| && \text{by (1.5.0.3).}\\
&\text{ iff } \mathfrak{A}' \models \psi\left[h'\binom{x}{a'}\right] && \text{for all } a' \in |\mathfrak{A}'| && \tau \text{ surjective } (\mathrm{I}_1)\\
&\text{ iff } \mathfrak{A}' \models \forall x\,\psi\,[h'] && && \text{by (1.5.0.9).}
\end{aligned}
$$

This completes the proof of (2.2.1.1).

To prove the final claim of the theorem, that $\mathfrak{A} \equiv \mathfrak{A}'$, recall that if $\varphi$ is a sentence, then the satisfaction of $\varphi$ in $\mathfrak{A}$ does not depend on the evaluation $h$ considered (1.5.1.3). Thus (2.2.1.1) implies that for any $L$-sentence $\varphi$, $\mathfrak{A} \models \varphi$ if and only if $\mathfrak{A}' \models \varphi$, which is the definition of $\mathfrak{A} \equiv \mathfrak{A}'$. □

We have thus proved that for $L$-structures $\mathfrak{A}$ and $\mathfrak{A}'$:

$$
\mathfrak{A} \cong \mathfrak{A}' \quad \text{implies} \quad \mathfrak{A} \equiv \mathfrak{A}'. \tag{2.2.1.3}
$$

A condition for the converse of (2.2.1.3) to hold will be given in Theorem 2.2.3 below. But first we wish to make a remark on the proof of Theorem 2.2.1.

**Remark 2.2.2** *Let $\tau : |\mathfrak{A}| \to |\mathfrak{A}'|$ be a mapping from $\mathfrak{A}$ into $\mathfrak{A}'$. If $\tau$ is a monomorphism, then for every quantifier-free formula $\varphi$ and every evaluation $h$ in $\mathfrak{A}$, we have:*

$$
\mathfrak{A} \models \varphi\,[h] \quad \textit{iff} \quad \mathfrak{A}' \models \varphi\,[\tau \circ h]. \tag{2.2.2.1}
$$

*Conversely, if* (2.2.2.1) *holds for all quantifier-free formulae* $\varphi$*, then* $\tau$ *is a monomorphism.*

*Proof*: Since, in the proof of Theorem 2.2.1, the surjectivity of $\tau$ was used only in the last case, where $\varphi$ was of the form $\forall x\,\psi$, it is clear that (2.2.2.1) holds for all formulae $\varphi$ in whose recursive construction no quantifier $\forall x$ is used. These are just the quantifier-free formulae.

Conversely, if we assume (2.2.2.1) for all quantifier-free $\varphi$, and apply it (with a suitable evaluation $h$) to the quantifier-free formulae

$$\begin{array}{ll} R_i(v_1,\ldots,v_{\lambda(i)}) & \text{for } i\in I,\\ f_j(v_1,\ldots,v_{\mu(j)}) \doteq v_0 & \text{for } j\in J, \text{ and}\\ c_k \doteq v_0 & \text{for } k\in K, \end{array}$$

then one obtains conditions ($I_2$), ($I_3'$), and ($I_4'$), respectively. Finally, applying it to the formula

$$v_0 \not\doteq v_1$$

gives the injectivity of $\tau$. □

**Theorem 2.2.3.** *The L-structures of finite cardinality, and only those, can be "characterized" up to L-isomorphism; i.e. there is a set* $\Sigma \subseteq \mathrm{Sent}(L)$ *such that for every L-structure* $\mathfrak{A}'$*:*

$$\mathfrak{A}' \models \Sigma \quad \textit{iff} \quad \mathfrak{A}' \cong \mathfrak{A}. \tag{2.2.3.1}$$

*Proof*: First assume that $\mathfrak{A}$ is infinite. If there were a set $\Sigma \subseteq \mathrm{Sent}(L)$ such that for all $\mathfrak{A}'$, (2.2.3.1) holds, then, in particular, $\Sigma \subseteq \mathrm{Th}(\mathfrak{A})$. Now, since $\mathfrak{A}$ is infinite, Corollary 2.1.2 implies that there is an elementarily equivalent $\mathfrak{A}'$ with some cardinality $\kappa > \mathrm{card}(|\mathfrak{A}|)$. Therefore $\mathfrak{A}' \models \mathrm{Th}(\mathfrak{A})$, and, *a fortiori*, $\mathfrak{A}' \models \Sigma$, but $\mathfrak{A}' \not\cong \mathfrak{A}$, disproving (2.2.3.1).

Second, assume $\mathfrak{A}$ is finite, i.e. the universe $|\mathfrak{A}|$ is finite – say, of cardinality $n$.

We begin by considering the case where the *index sets* $I$, $J$ and $K$ (1.2.0.1)–(1.2.0.3) of $L$ are likewise *finite*. In this case we shall find a single sentence $\alpha \in \mathrm{Th}(\mathfrak{A})$ such that for all $L$-structures $\mathfrak{A}'$,

$$\mathfrak{A}' \models \alpha \quad \text{implies} \quad \mathfrak{A} \cong \mathfrak{A}'. \tag{2.2.3.2}$$

Our sentence $\alpha$ will be the conjunction of two $L$-sentences, each of which holds in $\mathfrak{A}$. As the first conjunct, we wish to take a sentence that ensures that $|\mathfrak{A}'|$ has no more elements than $|\mathfrak{A}|$. For this, we use the cardinality-sentence $\alpha_{\leq n}$, defined as

$$\exists v_1,\ldots,v_n\, \forall v_0\, (v_0 \doteq v_1 \vee \cdots \vee v_0 \doteq v_n). \tag{2.2.3.3}$$

If $\alpha_{\leq n}$ holds in $\mathfrak{A}'$, then $|\mathfrak{A}'|$ obviously has at most $n$ elements. Now let $a_1,\ldots,a_n$ be the elements of $|\mathfrak{A}|$. Fix an evaluation $h$ in $\mathfrak{A}$ with $h(v_i) = a_i$ for $i = 1,2,\ldots,n$. We consider the set $\Phi$ of all formulae that, on the one hand, hold in $\mathfrak{A}$ under $h$, and, on the other hand, are of one of the following five types:

$$\begin{array}{lll} x_0 \neq x_1 & & \\ R_i(x_1,\dots,x_{\lambda(i)}) & \text{with} & i \in I \\ \neg R_i(x_1,\dots,x_{\lambda(i)}) & \text{with} & i \in I \\ f_j(x_1,\dots,x_{\mu(j)}) \doteq x_0 & \text{with} & j \in J \\ c_k \doteq x_0 & \text{with} & k \in K, \end{array}$$

for any $x_0, x_1, \ldots \in \{v_1, \ldots, v_n\}$. One sees easily that $\Phi$ is a finite set. Write $\bigwedge \Phi$ for the conjunction of all formulae in $\Phi$. We claim that the desired sentence $\alpha$ is then

$$(\alpha_{\leq n} \wedge \exists v_1, \ldots, v_n \bigwedge \Phi).$$

If $\alpha$ holds in $\mathfrak{A}'$, then, first, $|\mathfrak{A}'|$ has at most $n$ elements, and, second, there is an evaluation $h'$ in $\mathfrak{A}'$ such that for each $\varphi \in \Phi$, $\mathfrak{A}' \models \varphi[h']$, by (1.5.0.11) and (1.5.0.8). If we now set $\tau(a_i) = h'(v_i)$ for $i = 1, 2, \ldots, n$, then we get a mapping

$$\tau : |\mathfrak{A}| \to |\mathfrak{A}'|$$

such that for each formula $\varphi$ of one of the five types above,

$$\mathfrak{A} \models \varphi[h] \quad \text{implies} \quad \mathfrak{A}' \models \varphi[\tau \circ h], \tag{2.2.3.4}$$

since on the free variables of $\varphi$, the evaluations $h'$ and $\tau \circ h$ agree with each other. But (2.2.3.4) has as consequences exactly the injectivity of $\tau$ and the conditions ($I_2$), ($I'_3$), and ($I'_4$). Therefore $\tau$ is a monomorphism of $\mathfrak{A}$ into $\mathfrak{A}'$. And since $|\mathfrak{A}'|$ has at most $n$ elements, $\tau$ is even an isomorphism.

Finally, we consider the case of arbitrary index sets $I$, $J$ and $K$. We claim that we may take $\Sigma$ in (2.2.3.1) to be $\mathrm{Th}(\mathfrak{A})$. For this, assume $\mathfrak{A}'$ is a model of $\mathrm{Th}(\mathfrak{A})$. Then since $\alpha_{\leq n} \in \mathrm{Th}(\mathfrak{A})$, $|\mathfrak{A}'|$ has at most $n$ elements. And since

$$\exists x_1, \ldots, x_n \left( \bigwedge_{i<j} x_i \neq x_j \right) \in \mathrm{Th}(\mathfrak{A}),$$

$|\mathfrak{A}'|$ has at least $n$ elements. Thus there are exactly $n!$ bijections from $|\mathfrak{A}|$ onto $|\mathfrak{A}'|$; we denote them by $\tau_1, \ldots, \tau_{n!}$. Suppose, for the sake of a contradiction, that none of these bijections is an $L$-isomorphism. Then for every $\nu \leq n!$, there is at least one index $i \in I$ or $j \in J$ or $k \in K$ such that one of the conditions ($I_2$), ($I_3$), or ($I_4$) is violated for $\tau_\nu$. For each $\nu$, fix one such "troublesome index". Altogether we fix at most $n!$ troublesome indices; they lie in certain finite subsets $I_1$ of $I$, $J_1$ of $J$ and $K_1$ of $K$. These subsets determine a sublanguage $L_1$ of $L$. Let $\mathfrak{A}_1$ and $\mathfrak{A}'_1$ be the restrictions of the $L$-structures $\mathfrak{A}$ and $\mathfrak{A}'$, respectively, to the language $L_1$ (they result from $\mathfrak{A}$ and $\mathfrak{A}'$ simply by dropping the interpretations of $R_i$ for $i \in I \setminus I_1$, of $f_j$ for $j \in J \setminus J_1$ and of $c_k$ for $k \in K \setminus K_1$). By (2.2.3.2) there is an $L_1$-isomorphism $\tau : \mathfrak{A}_1 \leftrightarrow \mathfrak{A}'_1$. Since $|\mathfrak{A}| = |\mathfrak{A}_1|$ and $|\mathfrak{A}'| = |\mathfrak{A}'_1|$, $\tau$ must coincide with one of the mappings $\tau_\nu$. But then $\tau$ could not be an $L_1$-isomorphism of $\mathfrak{A}_1$ onto $\mathfrak{A}'_1$, since $\tau_\nu$'s troublesome index lies in

$I_1$, $J_1$ or $K_1$. This contradiction shows that (at least) one of the mappings $\tau_1, \ldots, \tau_{n!}$ must, indeed, be an $L$-isomorphism of $\mathfrak{A}$ onto $\mathfrak{A}'$. □

Now we would like to introduce a suggestive notation for $\mathfrak{A} \models \varphi[h]$, which makes use of the fact (Lemma 1.5.1(b)) that the satisfaction of a formula $\varphi$ in $\mathfrak{A}$ under an evaluation $h$ depends only on the values of $h$ at the free variables of $\varphi$. If $\mathrm{Fr}(\varphi) \subseteq \{v_0, \ldots, v_n\}$ and $a_0, \ldots, a_n \in |\mathfrak{A}|$, then we wish to write simply

$$\mathfrak{A} \models \varphi[a_0, \ldots, a_n]$$

for $\mathfrak{A} \models \varphi[h]$, where $h$ is any evaluation in $\mathfrak{A}$ with $a_\nu = h(v_\nu)$ for $\nu = 0, 1, \ldots, n$. And for $\mathfrak{A} \models \varphi\left[h\binom{x}{a}\right]$ we write

$$\mathfrak{A} \models \varphi\left[a_0, \ldots, a_n, \binom{x}{a}\right].$$

Using this notation, the conclusion (2.2.1.1) of Theorem 2.2.1 may now be expressed as follows: for every $L$-formula $\varphi$ with $\mathrm{Fr}(\varphi) \subseteq \{v_0, \ldots, v_n\}$ and all $a_0, \ldots,$ $a_n \in |\mathfrak{A}|$,

$$\mathfrak{A} \models \varphi[a_0, \ldots, a_n] \quad \text{iff} \quad \mathfrak{A}' \models \varphi[\tau(a_0), \ldots, \tau(a_n)].$$

## 2.3 Substructures

In this section we shall become acquainted with a "relative" construction method that goes back to Löwenheim and Skolem. It allows us to construct, from any infinite $L$-structure $\mathfrak{A}$, a so-called elementary substructure $\mathfrak{B}$ that satisfies certain secondary conditions. $\mathfrak{B}$ will, in particular, be elementarily equivalent to $\mathfrak{A}$. But first to the definitions of "substructure" and "elementary substructure"!

An $L$-structure $\mathfrak{B}$ is called a *substructure* of an $L$-structure $\mathfrak{A}$ if $|\mathfrak{B}| \subseteq |\mathfrak{A}|$ and the inclusion mapping $\mathrm{id} : |\mathfrak{B}| \to |\mathfrak{A}|$ is an embedding. This means that for $a_1, a_2, \ldots \in$ $|\mathfrak{B}|$ we have:

$$\begin{aligned} R_i^{\mathfrak{B}}(a_1, \ldots) &\text{ iff } R_i^{\mathfrak{A}}(a_1, \ldots) && \text{for} \quad i \in I \\ f_j^{\mathfrak{B}}(a_1, \ldots) &= f_j^{\mathfrak{A}}(a_1, \ldots) && \text{for} \quad j \in J \\ c_k^{\mathfrak{B}} &= c_k^{\mathfrak{A}} && \text{for} \quad k \in K. \end{aligned}$$

If $\mathfrak{B}$ is a substructure of $\mathfrak{A}$, then $\mathfrak{A}$ is also called an *extension structure* of $\mathfrak{B}$; we write $\mathfrak{B} \subseteq \mathfrak{A}$ for this.

For a substructure $\mathfrak{B}$ of $\mathfrak{A}$, the subset $B := |\mathfrak{B}|$ of $|\mathfrak{A}|$ has the following properties:

(i) $B$ is closed under the functions $f_j^{\mathfrak{A}}$;
i.e. for all $a_1, \ldots, a_{\mu(j)} \in B$, $f_j^{\mathfrak{B}}(a_1, \ldots, a_{\mu(j)}) \in B$;

(ii) the $\mathfrak{A}$-interpretations $c_k^{\mathfrak{A}}$ of the constant symbols lie in $B$; i.e. $c_k^{\mathfrak{A}} \in B$.

If a nonempty subset $B$ of $|\mathfrak{A}|$ satisfies these conditions, then the ordered quadruple

$$\mathfrak{B} := \langle B; (R_i^{\mathfrak{A}}|_B)_{i\in I}; (f_j^{\mathfrak{A}}|_B)_{j\in J}; (c_k^{\mathfrak{A}})_{k\in K}\rangle \tag{2.3.0.1}$$

is clearly also an $L$-structure, and indeed a substructure of $\mathfrak{A}$, called the substructure of $\mathfrak{A}$ *defined by B*. Here $R_i^{\mathfrak{A}}|_B$ and $f_j^{\mathfrak{A}}|_B$ are the restrictions to $B$ of the relations $R_i^{\mathfrak{A}}$ and the functions $f_j^{\mathfrak{A}}$.

Let us briefly consider an example. Let

$$\mathfrak{R} := \langle \mathbb{R}; +, \cdot, -; 0\rangle,$$

where $+$, $\cdot$, and $-$ are understood as (the usual) 2-place operations on $\mathbb{R}$, and 0 and 1 have their usual meaning. Then the subset of integers $\mathbb{Z} \subseteq \mathbb{R}$ defines, by (2.3.0.1), a substructure of $\mathfrak{R}$. The subset of natural numbers $\mathbb{N}$ does not define a substructure, since it is not closed under subtraction (i). The subset $2\mathbb{Z}$ of even integers likewise does not define a substructure, since it does not contain the number 1 (ii).

A substructure $\mathfrak{B}$ of $\mathfrak{A}$ is called an *elementary substructure* if for all $L$-formulae $\varphi$ and all evaluations $h$ in $\mathfrak{B}$, we have

$$\mathfrak{B} \models \varphi[h] \text{ iff } \mathfrak{A} \models \varphi[h]. \tag{2.3.0.2}$$

If $\mathfrak{B}$ is an elementary substructure of $\mathfrak{A}$, then we write $\mathfrak{B} \preceq \mathfrak{A}$. Obviously,

$$\mathfrak{B} \preceq \mathfrak{A} \quad \text{implies} \quad \mathfrak{B} \equiv \mathfrak{A}.$$

The converse of this does not hold, as the following example shows. Let $\mathbb{N}^+ = \mathbb{N}\setminus\{0\}$. Then

$$\langle \mathbb{N}^+; \leq\rangle \subseteq \langle \mathbb{N}; \leq\rangle,$$

where the two occurrences of the relation $\leq$ are to be understood on their corresponding domains. We have, in addition,

$$\langle \mathbb{N}^+; \leq\rangle \cong \langle \mathbb{N}; \leq\rangle,$$

where the isomorphism is given by $a \mapsto a-1$. Therefore, by Theorem 2.2.1,

$$\langle \mathbb{N}^+; \leq\rangle \equiv \langle \mathbb{N}; \leq\rangle.$$

The substructure $\langle \mathbb{N}^+; \leq\rangle$ of $\langle \mathbb{N}; \leq\rangle$ is, however, not an elementary substructure, since, under any evaluation $h$ with $h(v_0) = 1$, the formula $\forall v_1\ v_0 \leq v_1$ holds in $\langle \mathbb{N}^+; \leq\rangle$, but not in $\langle \mathbb{N}; \leq\rangle$.

The following sufficient condition for the elementary substructure relation easily yields

$$\langle \mathbb{Q}; <\rangle \preceq \langle \mathbb{R}; <\rangle.$$

**Lemma 2.3.1.** *Let $\mathfrak{B}$ be a substructure of $\mathfrak{A}$. If, to every finite set of elements $a_1,\ldots,a_m \in |\mathfrak{B}|$ and every $a \in |\mathfrak{A}|$ there is an automorphism $\tau$ of $\mathfrak{A}$ with $\tau(a_i) = a_i$ for $i = 1,2,\ldots,m$ and $\tau(a) \in |\mathfrak{B}|$, then $\mathfrak{B}$ is an elementary substructure of $\mathfrak{A}$.*

*Proof*: We show by induction on the recursive construction of formulae that (2.3.0.2) holds for all formulae $\varphi$ and all evaluations $h$ in $\mathfrak{B}$. This proof is completely analogous to that of Theorem 2.2.1, upon taking the $\mathfrak{A}$, $\mathfrak{A}'$, and $\tau : \mathfrak{A} \leftrightarrow \mathfrak{A}'$ in (2.2.1) to be $\mathfrak{B}$, $\mathfrak{B}$, and $\mathrm{id} : \mathfrak{B} \leftrightarrow \mathfrak{B}$ (so that $h' = \tau \circ h = h$ in (2.2.1)).[1] The only part of the induction step that is different here is the part where we assume that $\varphi$ is of the form $\forall x\, \psi$. We separate this part of the proof of (2.3.0.2) into two cases:

*Case 1*: $\mathfrak{A} \models \forall x\, \psi\,[h]$. Then we have $\mathfrak{A} \models \psi\left[h\binom{x}{a}\right]$ for all $a \in |\mathfrak{A}|$, by (1.5.0.9). For all $a \in |\mathfrak{B}|$, $h\binom{x}{a}$ is an evaluation in $\mathfrak{B}$; so, by the inductive hypothesis, $\mathfrak{B} \models \psi\left[h\binom{x}{a}\right]$. This gives $\mathfrak{B} \models \forall x\, \psi\,[h]$.

*Case 2*: $\mathfrak{A} \not\models \forall x\, \psi\,[h]$. Then there exists an $a \in |\mathfrak{A}|$ with

$$\mathfrak{A} \not\models \psi\left[h\binom{x}{a}\right]. \tag{2.3.1.1}$$

Let $\mathrm{Fr}(\forall x\, \psi) = \{x_1, \ldots, x_n\}$ (p. 12) be the set of all the free variables in $\forall x\, \psi$. We choose an automorphism $\tau$ of $\mathfrak{A}$ with $\tau(h(x_i)) = h(x_i)$ for $1 \leq i \leq n$ and $\tau(h\binom{x}{a}(x)) = \tau(a) \in |\mathfrak{B}|$. If we apply Theorem 2.2.1 to (2.3.1.1) and the automorphism $\tau$ of $\mathfrak{A}$, we obtain

$$\mathfrak{A} \not\models \psi\left[\tau \circ h\binom{x}{a}\right]. \tag{2.3.1.2}$$

Since $\tau \circ h$ and $h$ agree with each other on the free variables of $\forall x\, \psi$ (by the construction of $\tau$), $\tau \circ h\binom{x}{a} = (\tau \circ h)\binom{x}{\tau(a)}$ and $h\binom{x}{\tau(a)}$ agree with each other on the free variables of $\psi$. Applying Lemma 1.5.1(b) to (2.3.1.2), we therefore obtain $\mathfrak{A} \not\models \psi\left[h\binom{x}{\tau(a)}\right]$. The inductive hypothesis now gives $\mathfrak{B} \not\models \psi\left[h\binom{x}{\tau(a)}\right]$, which leads, finally, to $\mathfrak{B} \not\models \forall x\, \psi\,[h]$. □

*Remark 2.3.2.* A finite structure $\mathfrak{A}$ cannot have a proper elementary extension $\mathfrak{B}$, as every cardinality sentence $\alpha_{\leq n}$ carries over from $\mathfrak{A}$ to $\mathfrak{B}$.

Just as the concept of a substructure can be understood as a special case of the concept of an embedding, so also the concept of an elementary substructure can be understood as a special case of an elementary embedding. An embedding $\tau : \mathfrak{A} \to \mathfrak{A}'$ of $L$-structures is called an *elementary embedding* if for every formula $\varphi$ and every evaluation $h$ in $\mathfrak{A}$, we have

$$\mathfrak{A} \models \varphi\,[h] \quad \text{iff} \quad \mathfrak{A}' \models \varphi\,[\tau \circ h]. \tag{2.3.2.1}$$

Applied to $L$-sentences $\varphi$, this implies, in particular, $\mathfrak{A} \equiv \mathfrak{A}'$.

Now we come to the previously announced construction method of Löwenheim and Skolem.

**Theorem 2.3.3.** *Let $\mathfrak{A}$ be an L-structure with infinite cardinality, and $B_0$ a subset of $|\mathfrak{A}|$. Then there is an elementary substructure $\mathfrak{B}$ of $\mathfrak{A}$ with $B_0 \subseteq |\mathfrak{B}|$ and* $\mathrm{card}(|\mathfrak{B}|) \leq \max\{\kappa_L, \mathrm{card}(B_0)\}$. *If, in addition, $\kappa_L \leq \mathrm{card}(B_0)$, then* $\mathrm{card}(|\mathfrak{B}|) = \mathrm{card}(B_0)$.

---

[1] Note that the $\tau$ mentioned in the hypothesis of Lemma 2.3.1 plays no role at this stage of the proof; it will appear only in "Case 2" below.

*Proof*: If one examines the proof of Lemma 2.3.1, it becomes clear that one must construct the universe $B$ of $\mathfrak{B}$ so that, first, it defines a substructure of $\mathfrak{A}$, and, second, for every "existence claim" in $\mathfrak{A}$ (with parameters from $B$), an example can be found in $B$. This will be guaranteed by the following construction.

For each existence formula $\varphi$ of the form $\exists x\,\psi$,[2] let $n_\varphi$ be the highest index of a free variable of $\varphi$; if $\varphi$ is a sentence, we set $n_\varphi = -1$. We define an $(n_\varphi+1)$-place function $g_\varphi$ on $A = |\mathfrak{A}|$ in the following way: if $a_0,\ldots,a_{n_\varphi} \in A$ and if $h$ is an evaluation with $h(v_i) = a_i$ for $0 \le i \le n_\varphi$,

$$\text{let } g_\varphi(a_0,\ldots,a_{n_\varphi}) \text{ be an } a \in A \text{ with } \mathfrak{A} \models \psi\big[h\big(\tbinom{x}{a}\big)\big],$$

if there is such an $a$, i.e. if $\mathfrak{A} \models \exists x\,\psi[h]$. Otherwise, let

$$g_\varphi(a_0,\ldots,a_{n_\varphi}) = d_0,$$

where $d_0$ is an element of $A$ that is fixed for all existence formulae. Then we define

$$B = \bigcup_{m\in\mathbb{N}} B_m, \tag{2.3.3.1}$$

where we set

$$B_{m+1} = B_m \cup \bigcup_{\varphi} g_\varphi\big(B_m^{n_\varphi+1}\big). \tag{2.3.3.2}$$

The union in (2.3.3.2) runs over all existence formulae $\varphi$. And $D^{n_\varphi+1}$ denotes, as usual, the $(n_\varphi+1)$-fold Cartesian product of $D$; for $n_\varphi = -1$, $D^0 = \emptyset$. (A 0-place function from $D$ into $D$ consists, as usual, of a single element $d \in D$.)

We claim that the subset $B$ constructed in (2.3.3.1) defines a substructure of $\mathfrak{A}$. For this we must show that all $c_k^{\mathfrak{A}}$ lie in $B$, and $B$ is closed under all functions $f_j$. For $k \in K$ we choose as our existence formula $\varphi$ the formula $\exists v_0\; v_0 \doteq c_k$. There is an $a \in A$ with $\mathfrak{A} \models v_0 \doteq c_k\,\big[h\big(\tbinom{v_0}{a}\big)\big]$ for every evaluation $h$ – namely, $a = c_k^{\mathfrak{A}}$. Therefore the 0-place function $g_\varphi$ consists of such an $a$. However, since there is only one such $a$, $g_\varphi = c_k^{\mathfrak{A}}$. Therefore we have

$$c_k^{\mathfrak{A}} = g_\varphi \in B_1 \subseteq B.$$

For $j \in J$ we choose the existence formula $\varphi$ to be

$$\exists v_{\mu(j)}\; v_{\mu(j)} \doteq f_j(v_0,\ldots,v_{\mu(j)-1}).$$

The arity of the function $g_\varphi$ is then $n_\varphi + 1 = \mu(j)$. If $a_0,\ldots,a_{n_\varphi} \in B$, say $a_0,\ldots,a_{n_\varphi} \in B_m$, then we have, by construction,

$$a_{\mu(j)} = g_\varphi(a_0,\ldots,a_{n_\varphi}) \in B_{m+1} \subseteq B.$$

[2] Note that this notion differs from that of an "$\exists$-sentence" introduced Section 2.5.

Here $a_{\mu(j)}$ is an element $a$ in $A$ with $\mathfrak{A} \models v_{\mu(j)} \doteq f_j(v_0,\ldots,v_{\mu(j)-1})\left[h\binom{v_{\mu(j)}}{a}\right]$, where $h$ is an evaluation with $h(v_i) = a_i$ for $0 \leq i < \mu(j)$. There is, however, exactly one such $a$, namely, $f_j^{\mathfrak{A}}(a_0,\ldots,a_{\mu(j)-1})$. It therefore follows that

$$f_j^{\mathfrak{A}}(a_0,\ldots,a_{\mu(j)-1}) = g_\varphi(a_0,\ldots,a_{n_\varphi}) \in B.$$

We carry out the proof that the substructure $\mathfrak{B}$ is an elementary substructure of $\mathfrak{A}$ again by induction on the construction of formulae, analogously to the proof of Lemma 2.3.1. Here all that remains to show is that in the case of a "universal formula" $\forall x\,\psi$, its falsity (for an evaluation $h$ in $\mathfrak{B}$) transfers from $\mathfrak{A}$ to $\mathfrak{B}$. For this, suppose $\mathfrak{A} \not\models \forall x\,\psi\,[h]$. Then we have $\mathfrak{A} \models \varphi\,[h]$, where $\varphi$ is the existence formula $\exists x \neg\psi$. If, say,

$$h(v_0),\ldots,h(v_{n_\varphi}) \in B_m,$$

then, by construction,

$$a = g_\varphi(h(v_0),\ldots,h(v_{n_\varphi})) \in B_{m+1}$$

is an element of $B$ with $\mathfrak{A} \models \neg\psi\left[h\binom{x}{a}\right]$. Now since $h\binom{x}{a}$ is an evaluation in $\mathfrak{B}$, by the induction hypothesis we obtain $\mathfrak{B} \models \neg\psi\left[h\binom{x}{a}\right]$. Hence, in particular, $\mathfrak{B} \models \neg\psi\,[h]$. Thus the proof of $\mathfrak{B} \not\models \forall x\,\psi\,[h]$ is achieved.

It remains to give an upper bound on $\mathfrak{B}$. We show

$$\mathrm{card}(B_m) \leq \max\{\kappa_L, \mathrm{card}(B_0)\} = \kappa;$$

the claim then follows with the usual cardinality estimates. For $B_0$ this is clear. Let us now assume that the upper bound above holds for $B_m$. Then

$$\mathrm{card}(B_{m+1}) \leq \max\{\kappa, \mathrm{card}(\mathrm{Fml}(L))\} = \kappa,$$

since every set $g_\varphi\left(B_m^{n_\varphi+1}\right)$ of the union in (2.3.3.2), as well as $B_m$, has cardinality $\leq \kappa$, and the union is indexed with a subset of $\mathrm{Fml}(L)$. □

If one compares this construction, say, with the construction of the algebraic closure of a field $F$ in an extension field $L$ of $F$, one easily recognizes the similarity. In the field case we arrange, by adjunction of roots of polynomials with coefficients in $F$, that certain existence sentences that hold in $L$ hold also in the closure of $F$ in $L$ – namely, precisely those sentences that assert the existence of a root. In the case of the construction in Theorem 2.3.3, *all* existence sentences are considered.

**Corollary 2.3.4.** *Let $\kappa_L = \aleph_0$, i.e. let $L$ be a countable language. Then every $L$-structure $\mathfrak{A}$ with infinite cardinality possesses a countable elementary substructure $\mathfrak{B}$. Moreover, every elementary substructure $\mathfrak{B}$ of $\mathfrak{A}$ is infinite.*

*Proof*: We choose $B_0 = \emptyset$ in Theorem 2.3.3. Then $\mathrm{card}(|\mathfrak{B}|) \leq \aleph_0$. $|\mathfrak{B}|$, however, cannot be finite, since otherwise a cardinality-sentence $\alpha_{\leq n}$ (2.2.3.3) would hold in $\mathfrak{B}$, which naturally fails in $\mathfrak{A}$. □

## 2.4 Elementary Extensions and Chains

Let $\mathfrak{A}$ and $\mathfrak{B}$ be $L$-structures. $\mathfrak{B}$ is called an *elementary extension* of $\mathfrak{A}$ if $\mathfrak{A}$ is an elementary substructure of $\mathfrak{B}$. This not so profound definition will serve as a transition to another "relative" construction: from an $L$-structure $\mathfrak{A}$ we shall construct an elementary extension. We shall see that it is useful to first treat "constant extensions" of a language $L$ somewhat more systematically.

Let the language $L := (\lambda, \mu, K)$ be given (1.2.0.11). A *constant extension* of $L$ will be a language $L' = (\lambda, \mu, K')$, where $K' = K \cup \underline{K}$ and $K \cap \underline{K} = \emptyset$ (recall that such extensions were constructed in the proof of Theorem 1.4.2). The set $\underline{K}$ serves as an index set for "new" constant symbols. An $L'$-structure $\mathfrak{A}'$ is an $L$-structure $\mathfrak{A}$ together with an interpretation

$$\sigma : \underline{K} \to |\mathfrak{A}|$$

of the new constant symbols. For $k \in \underline{K}$, $c_k^{\mathfrak{A}'} = \sigma(k)$. For this we also write

$$\mathfrak{A}' = (\mathfrak{A}, \sigma);$$

we call $\mathfrak{A}$ the *restriction* of $\mathfrak{A}'$ to $L$.

The next lemma will prove to be very useful:

**Lemma 2.4.1.** *Let* $\mathfrak{A}' = (\mathfrak{A}, \sigma)$ *be an* $L'$*-structure, and* $\varphi$ *an* $L$*-formula with* $\mathrm{Fr}(\varphi) \subseteq \{v_0, \ldots, v_n\}$. *Then for all* $k_0, \ldots, k_n \in \underline{K}$ *and for all evaluations* $h$ *in* $\mathfrak{A}$ *with* $h(v_i) = \sigma(k_i)$ *for* $0 \le i \le n$,

$$\mathfrak{A} \models \varphi\,[h] \quad \textit{iff} \quad (\mathfrak{A}, \sigma) \models \varphi(v_0/c_{k_0}, \ldots, v_n/c_{k_n}). \qquad (2.4.1.1)$$

*Proof*: We prove this lemma by iterated application of Lemma 1.5.3 to $\mathfrak{A}'$. First, for $v_0$ we obtain, using $h(v_0) = c_{k_0}^{\mathfrak{A}'}[h]$:

$$\mathfrak{A}' \models \varphi\,[h] \quad \text{iff} \quad \mathfrak{A}' \models \varphi(v_0/c_{k_0})\,[h].$$

Iterating, we obtain, finally,

$$\mathfrak{A}' \models \varphi\,[h] \quad \text{iff} \quad \mathfrak{A}' \models \varphi(v_0/c_{k_0}, \ldots, v_n/c_{k_n})\,[h].$$

This, however, is equivalent to (2.4.1.1), since, on the one hand, $\varphi$ is an $L$-formula, and, on the other hand, $\varphi(v_0/c_{k_0}, \ldots, v_n/c_{k_n})$ is an $L'$-sentence. □

An especially useful special case of a constant extension $L'$ of $L$ presents itself when $L'$ is constructed with reference to a given $L$-structure $\mathfrak{A}$, as follows: suppose $A'$ is a subset of $|\mathfrak{A}|$. We assume that $K \cap |\mathfrak{A}| = \emptyset$, and we set $\underline{K} = A'$. In this case we write $\underline{a} := c_a$ for $a \in \underline{K}$. The canonical interpretation in $\mathfrak{A}$ of the new constant symbol $\underline{a}$ is, then, $a$ itself; that is, we use $\underline{a}$ as a name for $a$. The canonical $L'$-structure corresponding to $\mathfrak{A}$ is then

$$(\mathfrak{A}, A') := (\mathfrak{A}, \mathrm{id}_{A'}).$$

In order to make explicit the dependence of this constant extension on $A'$, we also write $L(A')$ for $L'$ in this case. If $A' = |\mathfrak{A}|$, then every element $a \in |\mathfrak{A}|$ possesses at least the name $\underline{a}$ in the language $L(|\mathfrak{A}|)$. In this case, we write, briefly, $L(\mathfrak{A})$ for $L(|\mathfrak{A}|)$. Then (2.4.1.1) can now be written as follows:

$$\mathfrak{A} \models \varphi[a_0,\ldots,a_n] \quad \text{iff} \quad (\mathfrak{A},|\mathfrak{A}|) \models \varphi(v_0/\underline{a_0},\ldots,v_n/\underline{a_n}).$$

If $\mathfrak{A}$ is an $L$-structure, then we define the *diagram of* $\mathfrak{A}$ to be the set $D(\mathfrak{A})$ consisting of those atomic $L(\mathfrak{A})$-sentences and negated atomic $L(\mathfrak{A})$-sentences that hold in $(\mathfrak{A},|\mathfrak{A}|)$. The diagram of $\mathfrak{A}$ is, therefore, a subset of the theory of $(\mathfrak{A},|\mathfrak{A}|)$ (2.1.1.7), the latter consisting precisely of all $L(\mathfrak{A})$-sentences that hold in $(\mathfrak{A},|\mathfrak{A}|)$.

**Lemma 2.4.2** (Diagram Lemma). *Let* $\mathfrak{A}$ *be an L-structure, and* $(\mathfrak{B},\sigma)$ *an* $L(\mathfrak{A})$*-structure. If* $(\mathfrak{B},\sigma)$ *is a model of* $D(\mathfrak{A})$*, then* $\sigma$ *is an L-embedding of* $\mathfrak{A}$ *into* $\mathfrak{B}$*. If* $(\mathfrak{B},\sigma)$ *is a model of* $\mathrm{Th}(\mathfrak{A},|\mathfrak{A}|)$*, then* $\sigma$ *is an elementary L-embedding of* $\mathfrak{A}$ *into* $\mathfrak{B}$*.*

*Proof*: By the definition of an embedding (recall the beginning of Section 2.2), we must show that the mapping $\sigma : |\mathfrak{A}| \to |\mathfrak{B}|$ satisfies the conditions $(\mathrm{I}_2)$, $(\mathrm{I}_3')$ and $(\mathrm{I}_4')$ there, and is, moreover, injective.

The proof of the injectivity of $\sigma$ goes as follows: if $a_1, a_2 \in |\mathfrak{A}|$ are unequal, then the sentence $\underline{a_1} \neq \underline{a_2}$ belongs to the diagram $D(\mathfrak{A})$. Since $(\mathfrak{B},\sigma)$ is a model of $D(\mathfrak{A})$, we have

$$(\mathfrak{B},\sigma) \models \underline{a_1} \neq \underline{a_2}.$$

If one observes that the interpretation of $\underline{a_i}$ in $(\mathfrak{B},\sigma)$ is just $\sigma(a_i)$, it follows that $\sigma(a_1) \neq \sigma(a_2)$.

Now the proofs of $(\mathrm{I}_3')$ and $(\mathrm{I}_4')$ go completely analogously, by considering the sentences $f_j(\underline{a_1},\ldots,\underline{a_{\mu(j)}}) \doteq \underline{d}$ and $c_k \doteq \underline{d}$, respectively. For $(\mathrm{I}_2)$, we utilize the sentence $R_i(\underline{a_1},\ldots,\underline{a_{\lambda(i)}})$ if $R_i^{\mathfrak{A}}(a_1,\ldots,a_{\lambda(i)})$ holds, and the sentence $\neg R_i(\underline{a_1},\ldots,\underline{a_{\lambda(i)}})$ otherwise; together we then obtain:

$$R_i^{\mathfrak{A}}(a_1,\ldots,a_{\lambda(i)}) \quad \text{iff} \quad R_i^{\mathfrak{B}}(\sigma(a_1),\ldots,\sigma(a_{\lambda(i)})).$$

Therefore $\sigma : |\mathfrak{A}| \to |\mathfrak{B}|$ is an embedding if $(\mathfrak{B},\sigma)$ is a model of $D(\mathfrak{A})$.

If $(\mathfrak{B},\sigma)$ is a model of $\mathrm{Th}(\mathfrak{A},|\mathfrak{A}|)$, then it is, in particular, a model of $D(\mathfrak{A})$. Therefore $\sigma$ is again an embedding. In order to show that $\sigma$ is even an elementary embedding of $\mathfrak{A}$ into $\mathfrak{B}$, we consider an $L$-formula $\varphi$ with $\mathrm{Fr}(\varphi) \subseteq \{v_0,\ldots,v_n\}$, and an evaluation $h$ in $\mathfrak{A}$. We must then show:

$$\mathfrak{A} \models \varphi[h] \quad \text{iff} \quad \mathfrak{B} \models \varphi[\sigma \circ h].$$

For this it suffices to prove

$$\mathfrak{A} \models \varphi[h] \text{ implies } \mathfrak{B} \models \varphi[\sigma \circ h], \tag{2.4.2.1}$$

since we could then apply (2.4.2.1) to $\neg\varphi$, which would furnish us with the converse of (2.4.2.1).

We set $a_i = h(v_i)$ for $0 \leq i \leq n$. From $\mathfrak{A} \models \varphi[h]$ we then obtain $(\mathfrak{A}, |\mathfrak{A}|) \models \varphi(v_0/\underline{a_0}, \ldots, v_n/\underline{a_n})$, by Lemma 2.4.1. Thus the sentence $\varphi(v_0/\underline{a_0}, \ldots, v_n/\underline{a_n})$ is an element of $\mathrm{Th}(\mathfrak{A}, |\mathfrak{A}|)$, and thus holds in $(\mathfrak{B}, \sigma)$. If we now apply Lemma 2.4.1 to the evaluation $\sigma \circ h$ in $\mathfrak{B}$, we finally obtain $\mathfrak{B} \models \varphi[\sigma \circ h]$, since $(\sigma \circ h)(v_i) = \sigma(a_i)$. □

If now $\mathfrak{A}$ is an $L$-structure of infinite cardinality, then we can apply Theorem 2.1.1 to the set of $L(\mathfrak{A})$-sentences $\Sigma = \mathrm{Th}(\mathfrak{A}, |\mathfrak{A}|)$, and obtain thereby a model $(\mathfrak{B}, \sigma)$ of $\Sigma$ with cardinality $\kappa$, where $\kappa$ is a prescribed cardinal number $\geq \kappa_{L(\mathfrak{A})} = \max\{\kappa_L, \mathrm{card}(|\mathfrak{A}|)\}$. By Lemma 2.4.2, $\sigma$ is an elementary embedding of $\mathfrak{A}$ into $\mathfrak{B}$. If we identify $\mathfrak{A}$ with its $\sigma$-image in $\mathfrak{B}$, we have thus proved:

**Theorem 2.4.3.** *Let $\mathfrak{A}$ be an L-structure of infinite cardinality. Then for every cardinal number $\kappa \geq \max\{\kappa_L, \mathrm{card}(|\mathfrak{A}|)\}$, there exists an elementary extension $\mathfrak{B}$ of $\mathfrak{A}$ with cardinality $\kappa$.* □

Besides this consequence of the second part of Lemma 2.4.2, we would like to record two additional consequences of Lemma 2.4.2 that are very important for the applications. First a consequence of the second part:

**Corollary 2.4.4.** *Let $\mathfrak{A}$ and $\mathfrak{B}$ be L-structures, and $\tau : \mathfrak{A} \to \mathfrak{B}$ an embedding. The embedding $\tau$ is elementary if and only if $(\mathfrak{A}, \mathrm{id}_{|\mathfrak{A}|}) \equiv (\mathfrak{B}, \tau)$. In particular, $\mathfrak{A} \preceq \mathfrak{B}$ if and only if $(\mathfrak{A}, |\mathfrak{A}|) \equiv (\mathfrak{B}, |\mathfrak{A}|)$.*

*Proof*: Applying the second part of Lemma 2.4.2 to $\sigma = \tau$, we get that $\tau$ is an elementary embedding if $(\mathfrak{B}, \sigma)$ is a model of $\mathrm{Th}(\mathfrak{A}, |\mathfrak{A}|)$. This last is equivalent to $(\mathfrak{A}, \mathrm{id}_{|\mathfrak{A}|}) \equiv (\mathfrak{B}, \tau)$.

Now, conversely, let $\tau$ be an elementary embedding, and $\varphi'$ an $L(\mathfrak{A})$-sentence. We can think of $\varphi'$ as arising from an $L$-formula $\varphi$ by substitution of constant symbols $\underline{a_0}, \ldots, \underline{a_n}$ for the variables $v_0, \ldots, v_n$:

$$\varphi' = \varphi(v_0/\underline{a_0}, \ldots, v_n/\underline{a_n}).$$

Of course, for this we use only variables $v_i$ that otherwise do not occur in $\varphi'$, in order to perform the substitution without problems. Now if we have $(\mathfrak{A}, \mathrm{id}_{|\mathfrak{A}|}) \models \varphi'$, then by Lemma 2.4.1, we get $\mathfrak{A} \models \varphi[h]$ for every evaluation $h$ with $h(v_i) = a_i$. Since the embedding $\tau$ is, by hypothesis, elementary, we have $\mathfrak{B} \models \varphi[\tau \circ h]$; using Lemma 2.4.1 again we get, finally, $(\mathfrak{B}, \tau) \models \varphi'$. Since this holds for all $L(\mathfrak{A})$-sentences $\varphi'$, we obtain

$$(\mathfrak{A}, \mathrm{id}_{|\mathfrak{A}|}) \equiv (\mathfrak{B}, \tau).$$ □

Next we would like to derive a consequence of the first part of Lemma 2.4.2. For this we introduce briefly the substructure generated by a subset $A'$ of an $L$-structure $\mathfrak{A}$.

It is easy to see that the intersection of arbitrarily many substructures of $\mathfrak{A}$ is again a substructure of $\mathfrak{A}$ (unless it is empty). The universe of such an intersection is defined to be just the intersection of the universes of all those substructures. On this intersection (of the universes) we then consider the common restriction of all relations and functions (under which, naturally, this intersection is closed). The interpretation $c_k^{\mathfrak{A}}$ of each constant symbol lies in every substructure of $\mathfrak{A}$, and hence in our intersection. If $\emptyset \neq A' \subseteq |\mathfrak{A}|$, we define the *substructure of* $\mathfrak{A}$ *generated by* $A'$ to be the intersection of all substructures of $\mathfrak{A}$ whose universes include $A'$. Here we can also allow $A'$ to be empty provided the language contains at least one constant symbol. A *finitely generated substructure* of $\mathfrak{A}$ is a substructure that is generated by a finite subset $A'$ of $|\mathfrak{A}|$.

As a corollary to the first part of Lemma 2.4.2, we now have:

**Corollary 2.4.5.** *Let* $\mathfrak{A}$ *be an* $L$*-structure, and* $\Sigma \subseteq \mathrm{Sent}(L)$. *If every finitely generated substructure of* $\mathfrak{A}$ *is embeddable in a model of* $\Sigma$, *then* $\mathfrak{A}$ *is embeddable in a model of* $\Sigma$.

*Proof*: It suffices to show that the set $\Sigma \cup D(\mathfrak{A})$ of $L(\mathfrak{A})$-sentences has a model $(\mathfrak{B}, \sigma)$, for then, first, $\mathfrak{B}$ will be a model of $\Sigma$, and, second, $\sigma$ will be an embedding of $\mathfrak{A}$ in $\mathfrak{B}$, by Lemma 2.4.2. By the Finiteness Theorem (1.5.6), it will now, in turn, suffice to show that every finite subset $\Pi$ of $\Sigma \cup D(\mathfrak{A})$ has a model.

Such a set $\Pi$ can contain only finitely many sentences $\delta_1, \ldots, \delta_n$ of the diagram $D(\mathfrak{A})$. Let the "new" constant symbols (i.e. the constant symbols not already in $L$) occurring in the $\delta_i$ be $\underline{a_1}, \ldots, \underline{a_m}$, where each $a_j \in |\mathfrak{A}|$. Then the substructure $\mathfrak{A}'$ of $\mathfrak{A}$ generated by $A' := \{a_1, \ldots, a_m\}$ is a model of $\{\delta_1, \ldots, \delta_n\}$. By hypothesis there is a model $\mathfrak{B}$ of $\Sigma$ and an embedding of $\mathfrak{A}'$ into $\mathfrak{B}$. After identification, $\mathfrak{B}$ is therefore a model of $\Sigma \cup \{\delta_1, \ldots, \delta_n\}$, and hence, in particular, of $\Pi$. □

A further possible construction for elementary extensions, of which we shall make extensive use in the next section, is based upon the concept of an elementary chain of $L$-structures. We call a sequence $(\mathfrak{A}_n)_{n\in\mathbb{N}}$ of $L$-structures an *elementary chain* if $\mathfrak{A}_{n+1}$ is an elementary extension of $\mathfrak{A}_n$, for each $n \in \mathbb{N}$. We shall show that the soon-to-be-defined union structure $\bigcup_{n\in\mathbb{N}} \mathfrak{A}_n$ is an elementary extension of each $\mathfrak{A}_n$. In a generalization of this situation, we shall consider not only sequences of length $\omega$, with order-type being that of the natural numbers, but also sequences of length $\alpha$, where $\alpha$ is an arbitrary ordinal number.

We call a sequence $(\mathfrak{A}_\nu)_{\nu<\alpha}$ of $L$-structures an $\alpha$*-chain* if $\mathfrak{A}_\nu \subseteq \mathfrak{A}_\mu$ holds for $\nu < \mu < \alpha$. The *union structure* of the chain $(\mathfrak{A}_\nu)_{\nu<\alpha}$ is the $L$-structure $\mathfrak{A}$ defined as follows: we set $|\mathfrak{A}| = \bigcup_{\nu<\alpha} |\mathfrak{A}_\nu|$, and for $i \in I$, $j \in J$, $k \in K$, and all $a_1, a_2, \ldots \in |\mathfrak{A}_\nu|$, we define:

$$R_i^{\mathfrak{A}}(a_1, \ldots, a_{\lambda(i)}) \Leftrightarrow R_i^{\mathfrak{A}_\nu}(a_1, \ldots, a_{\lambda(i)})$$
$$f_j^{\mathfrak{A}}(a_1, \ldots, a_{\mu(j)}) = f_j^{\mathfrak{A}_\nu}(a_1, \ldots, a_{\mu(j)})$$
$$c_k^{\mathfrak{A}} = c_k^{\mathfrak{A}_\nu}.$$

In view of $\mathfrak{A}_\nu \subseteq \mathfrak{A}_\mu$ for $\mu < \nu$ and $\mu < \alpha$, this is actually a correct definition, i.e. it is independent of the index $\nu$ for which all the $a_1, a_2, \ldots$ lie in $|\mathfrak{A}_\nu|$. We also write

$$\bigcup_{\nu<\alpha} \mathfrak{A}_\nu$$

for the union structure $\mathfrak{A}$. Observe that in the case of a successor ordinal number $\alpha = \beta + 1$, we have:

$$\bigcup_{\nu<\alpha} \mathfrak{A}_\nu = \mathfrak{A}_\beta; \qquad (2.4.5.1)$$

i.e. in this case the union is exactly equal to the maximum member of the chain.

**Theorem 2.4.6.** *Let $\alpha$ be an ordinal number, and $(\mathfrak{A}_\nu)_{\nu<\alpha}$ be an $\alpha$-chain of L-structures. Suppose $\mathfrak{A}_\beta \preceq \mathfrak{A}_{\beta+1}$ for $\beta+1<\alpha$, and $\mathfrak{A}_\lambda = \bigcup_{\nu<\lambda} \mathfrak{A}_\nu$ for limit ordinal numbers $\lambda < \alpha$. Then the union structure $\mathfrak{A} = \bigcup_{\nu<\alpha} \mathfrak{A}_\nu$ is an elementary extension of $\mathfrak{A}_\nu$ for each $\nu < \alpha$. In particular, $\mathfrak{A}_0 \preceq \mathfrak{A}$.*

Observe that for the case of an $\omega$-chain $(\mathfrak{A}_n)_{n\in\mathbb{N}}$, the hypothesis for the case of limit ordinals is vacuously satisfied. Only the condition $\mathfrak{A}_n \preceq \mathfrak{A}_{n+1}$ remains.

*Proof*: We carry out an ordinal-induction proof on the length $\alpha$ of the chain. There are three cases to distinguish.

<u>$\alpha = 0$</u>: In this case both the chain and the claim are empty.

<u>$\alpha = \beta+1$</u>: In this case the chain $(\mathfrak{A}_\nu)_{\nu<\beta}$ is of shorter length, and still satisfies the hypotheses of the theorem. By the inductive hypothesis, then, $\mathfrak{A}_\mu \preceq \bigcup_{\nu<\beta} \mathfrak{A}_\nu$ for all $\mu < \beta$. It therefore remains to show that $\bigcup_{\nu<\beta} \mathfrak{A}_\nu \preceq \mathfrak{A}_\beta$ $(= \bigcup_{\nu<\alpha} \mathfrak{A}_\nu$, by (2.4.5.1)), since, by the obvious transitivity of the elementary substructure relation, $\mathfrak{A}_\mu \preceq \mathfrak{A}_\beta$ will then also hold. If $\beta$ is a limit ordinal, then $\bigcup_{\nu<\beta} \mathfrak{A}_\nu = \mathfrak{A}_\beta$ by hypothesis. If $\beta = \gamma+1$, then $\bigcup_{\nu<\beta} \mathfrak{A}_\nu = \mathfrak{A}_\gamma$ (2.4.5.1), and since $\gamma+1<\alpha$, $\mathfrak{A}_\gamma \preceq \mathfrak{A}_{\gamma+1} = \mathfrak{A}_\beta$, by hypothesis.

<u>$\alpha$ is a limit ordinal</u>: By the inductive hypothesis, we have $\mathfrak{A}_\mu \preceq \mathfrak{A}_\nu$ for $\mu < \nu < \alpha$. We must show $\mathfrak{A}_\mu \preceq \mathfrak{A}$ for $\mu < \alpha$. We prove this by proving, via induction on the construction of $\varphi \in \mathrm{Sent}(L)$, that for all $\mu < \alpha$ and all evaluations $h$ in $\mathfrak{A}_\mu$:

$$\mathfrak{A}_\mu \models \varphi[h] \quad \text{iff} \quad \mathfrak{A} \models \varphi[h].$$

As in the corresponding proofs in Section 2.3, this induction is purely routine up until the case of the transfer of the falsity of a formula $\forall x\, \psi$ from $\mathfrak{A}$ to $\mathfrak{A}_\mu$ under the evaluation $h$. So let $\mathfrak{A} \not\models \forall x\, \psi[h]$. Then there is an $a \in |\mathfrak{A}|$ with $\mathfrak{A} \not\models \psi\left[\binom{x}{a}\right]$. Let $\mathrm{Fr}(\psi) \subseteq \{v_0, \dots, v_n\}$ and $a_i = h(v_i)$ for $0 \le i \le n$. Since $|\mathfrak{A}|$ is an increasing union, $a, a_0, \dots, a_n$ lie already in a member $|\mathfrak{A}_\nu|$ of this union, where we may assume, without loss of generality, that $\mu \le \nu$. For the simpler formula $\psi$ and the evaluation $h\binom{x}{a}$ in $\mathfrak{A}_\nu$, it follows from the inductive hypothesis that $\mathfrak{A}_\nu \not\models \psi\left[h\binom{x}{a}\right]$. In particular, this gives $\mathfrak{A}_\nu \not\models \forall x\, \psi[h]$. Using $\mathfrak{A}_\mu \preceq \mathfrak{A}_\nu$ we obtain, finally, $\mathfrak{A}_\mu \not\models \forall x\, \psi[h]$. □

We call an $\alpha$-chain satisfying the hypothesis of Theorem 2.4.6 an *elementary $\alpha$-chain*. If an $\alpha$-chain is not elementary, then the truth of a sentence in the members of the chain does not in general transfer to the union. For especially simple sentences, however, this is right.

**Remark 2.4.7** *Let $\alpha$ be an ordinal number, and $(\mathfrak{A}_\nu)_{\nu<\alpha}$ be an $\alpha$-chain of L-structures. If an $\forall\exists$-sentence $\varphi$ holds in each member $\mathfrak{A}_\nu$ of the chain, then it holds also in $\bigcup_{\nu<\alpha} \mathfrak{A}_\nu = \mathfrak{A}$.*

Here, by an $\forall\exists$-*sentence* we mean a sentence of the form $\forall x_1,\ldots,x_n \exists y_1,\ldots,y_m \psi$, where $\psi$ is quantifier-free.

*Proof*: Let $\varphi$ be of the given form. Further, let $h$ be an evaluation in $\bigcup_{\nu<\alpha}\mathfrak{A}_\nu$, and $a_1,\ldots,a_n \in \bigcup_{\nu<\alpha}|\mathfrak{A}_\nu|$, so that $a_1,\ldots,a_n \in |\mathfrak{A}_\nu|$ for a certain $\nu < \alpha$. Let $h'$ be an evaluation in $\mathfrak{A}_\nu$ with $h'(x_i) = a_i$ for $1 \le i \le n$. By hypothesis we have $\mathfrak{A}_\nu \models \exists y_1,\ldots,y_m \psi[h']$. So, there exist $d_1,\ldots,d_m \in |\mathfrak{A}_\nu|$ with

$$\mathfrak{A}_\nu \models \psi\left[h'\binom{y_1}{d_1}\cdots\binom{y_m}{d_m}\right].$$

Since $\psi$ is quantifier-free, we can replace $\mathfrak{A}_\nu$ by $\mathfrak{A}$ here, and obtain thereby $\mathfrak{A} \models \exists y_1,\ldots,y_m \psi[h']$. By (1.5.1)(b), this is equivalent to

$$\mathfrak{A} \models \exists y_1,\ldots,y_m \psi\left[h\binom{x_1}{a_1}\cdots\binom{x_n}{a_n}\right].$$

Since $a_1,\ldots,a_n$ were chosen arbitrarily in $\bigcup_{\nu<\alpha}|\mathfrak{A}_\nu|$, we obtain, finally,

$$\mathfrak{A} \models \forall x_1,\ldots,x_n \exists y_1,\ldots,y_m \psi[h]. \qquad \square$$

*Example 2.4.8* (of the failure of Remark 2.4.7 for $\exists\forall$-sentences). To show the limitations of the statement made in Remark 2.4.7, we consider the following example. Let

$$\mathfrak{A}_n = \left\langle \frac{1}{n!}\mathbb{Z}; <_n \right\rangle.$$

Here, for $n \in \mathbb{N}$,

$$\frac{1}{n!}\mathbb{Z} = \left\{ \frac{m}{n!} \,\middle|\, m \in \mathbb{Z} \right\},$$

and $<_n$ is the restriction to this set of the usual ordering on $\mathbb{Q}$. It is clear that $(\mathfrak{A}_n)_{n\in\mathbb{N}}$ is an $\omega$-chain. The "$\exists\forall$-sentence"

$$\exists x \forall y (0 < x \wedge (y < 0 \vee x = y \vee x < y)) \tag{2.4.8.1}$$

holds in each member $\mathfrak{A}_n$ of the chain. Namely, $\frac{1}{n!}$ is a smallest positive element $x$ in $\frac{1}{n!}\mathbb{Z}$. But (2.4.8.1) obviously no longer holds in

$$\bigcup_{n<\omega}\mathfrak{A}_n = \langle \mathbb{Q}; <^{\mathbb{Q}} \rangle.$$

## 2.5 Saturated Structures

In this section we wish to introduce a concept that is fundamental to model theory: the concept of a saturated structure. The existence (and in a certain sense uniqueness) of saturated elementary extensions of a given, infinite structure is very helpful for the investigation of the theory. In the investigations in Chapter 3, we shall use saturated structures only very rarely (the simplicity of the theories considered there makes this unnecessary); however, we especially wish to point out their usefulness in the investigation of, say, valued fields (see Chapter 4). At the end of this section we wish to discuss several examples of saturated structures.

Let $\mathfrak{A}$ be an $L$-structure, where, as always, $L = (\lambda, \mu, K)$ is a given language. We denote the cardinality of $L$ again by $\kappa_L$ (cf. (2.1)). As described in (2.4), we extend $L$ by the addition of a new constant symbol $\underline{a}$ for each $a \in A'$, where $A'$ is some subset of $A := |\mathfrak{A}|$. We had denoted the resulting language by $L(A')$. If $\Phi$ is a set of formulae, the notation $\Phi(v_0)$ will be used only when $\mathrm{Fr}(\varphi) \subseteq \{v_0\}$ for all $\varphi \in \Phi$. We call a set $\Phi(v_0)$ of $L(A')$-formulae a(n *elementary*) *type* of $\mathfrak{A}$, or, more precisely, of $(\mathfrak{A}, A')$, if there is an elementary extension $\mathfrak{A}_1$ of $\mathfrak{A}$ and an $a \in A_1 := |\mathfrak{A}_1|$ with $(\mathfrak{A}_1, A_1) \models \Phi(\underline{a})$ (i.e. $(\mathfrak{A}_1, A_1) \models \varphi(\underline{a})$ for all $\varphi \in \Phi$). We say, then, that $\Phi(v_0)$ is *realized* or *satisfied* in $(\mathfrak{A}_1, A_1)$ by $a$.

From now on, instead of writing $\psi(x_1/c_1, \ldots, x_n/c_n)$, we shall always write, more briefly, $\psi(c_1, \ldots, c_n)$, when it is clear which variables are being replaced by the given constants.

**Lemma 2.5.1.** *A set $\Phi(v_0)$ of $L(A)$-formulae is a type of $\mathfrak{A}$ if and only if every finite subset $\{\varphi_1, \ldots, \varphi_n\}$ of $\Phi(v_0)$ is realizable in $(\mathfrak{A}, A)$, i.e. $(\mathfrak{A}, A) \models \exists v_0 (\varphi_1 \wedge \cdots \wedge \varphi_n)$.*

*Proof*: If $\Phi(v_0)$ is a type of $\mathfrak{A}$ and, say, $\mathfrak{A} \preceq \mathfrak{A}_1$ with $(\mathfrak{A}_1, A_1) \models \Phi(\underline{a})$ for some $a \in A_1$, then naturally, for all $\varphi_1, \ldots, \varphi_n \in \Phi$, $(\mathfrak{A}_1, A) \models \exists v_0 (\varphi_1 \wedge \cdots \wedge \varphi_n)$ as well, and hence (by elementariness of the extension) $(\mathfrak{A}, A) \models \exists v_0 (\varphi_1 \wedge \cdots \wedge \varphi_n)$.

Conversely, suppose every finite subset of $\Phi(v_0)$ is realizable in $(\mathfrak{A}, A)$. Then we consider the set

$$\Sigma = \mathrm{Th}(\mathfrak{A}, A) \cup \Phi(c),$$

of sentences, where $c$ is a new constant symbol. A model of $\Sigma$ would obviously furnish an elementary extension $\mathfrak{A}_1$ of $\mathfrak{A}$ by Lemma 2.4.2, and an interpretation $a = c^{\mathfrak{A}_1}$ of $c$, so that we thereby obtain $(\mathfrak{A}_1, A \cup \{a\}) \models \Phi(c)$. This would imply, trivially, that $(\mathfrak{A}_1, A_1) \models \Phi(\underline{a})$.

To show that $\Sigma$ has a model, it suffices, by the Finiteness Theorem 1.5.6, to show that every finite subset of $\Sigma$ has a model. Such a finite subset is, however, contained in a set of the form

$$\mathrm{Th}(\mathfrak{A}, A) \cup \{\varphi_1(c), \ldots, \varphi_n(c)\}, \tag{2.5.1.1}$$

for certain $\varphi_1, \ldots, \varphi_n \in \Phi(c)$. In view of $(\mathfrak{A}, A) \models \exists v_0 (\varphi_1 \wedge \cdots \wedge \varphi_n)$, however, (2.5.1.1) has a model, whence the given, finite subset of $\Phi(v_0)$ has one, too. □

Even when every finite subset of a type $\Phi(v_0)$ of $\mathfrak{A}$ is satisfiable in $(\mathfrak{A}, A)$, this need not in general hold for $\Phi(v_0)$ itself. The simplest counterexample is probably the following. Suppose $\mathfrak{A}$ is an infinite structure, and

$$\Phi(v_0) = \{v_0 \neq \underline{a} \mid a \in A\}. \tag{2.5.1.2}$$

By the infinitude of $A$, every finite subset of $\Phi(v_0)$ is satisfiable in $(\mathfrak{A}, A)$, while $\Phi(v_0)$ itself is not.

In a "saturated" structure $\mathfrak{A}$ we shall realize as many types of $\mathfrak{A}$ as possible. For an infinite cardinal number $\kappa$, we call an $L$-structure $\mathfrak{A}$ *$\kappa$-saturated* if every type $\Phi(v_0)$ of $(\mathfrak{A},A')$ with $\text{card}(A') < \kappa$ can be realized in $(\mathfrak{A},A)$. By Remark 2.3.2, a finite structure $\mathfrak{A}$ is always $\kappa$-saturated. For infinite $A$ we obtain immediately from $\kappa$-saturation that

$$\kappa \leq \text{card}(A); \tag{2.5.1.3}$$

otherwise, the type (2.5.1.2) would have to be realizable in $(\mathfrak{A},A)$. A further consequence is that each $\kappa$-saturated structure $\mathfrak{A}$ is also $\kappa'$-saturated for every infinite cardinal number $\kappa' \leq \kappa$. Furthermore, the concept of $\kappa$-saturation is obviously invariant under extensions of $L$ by fewer than $\kappa$ constant symbols. This is clear, since types of $\mathfrak{A}$ are sets of formulae in the language $L(A)$. New constant symbols can always be replaced with constant symbols $\underline{a}$ of the language $L(A)$, by (2.5.1.3). (Observe that an element $a \in A$ can be denoted not only by $\underline{a}$, but by additional constant symbols, as well, naturally.)

We shall now show that to every infinite structure, there are always sufficiently saturated elementary extensions. For a cardinal number $\kappa$, we denote the successor cardinal by $\kappa^+$; thus, $\kappa^+ \leq 2^\kappa$.

**Theorem 2.5.2** (Existence Theorem). *To every cardinal number $\kappa \geq \kappa_L$ and every infinite L-structure $\mathfrak{A}$ with* $\text{card}(A) \leq 2^\kappa$, *there is a $\kappa^+$-saturated elementary extension $\mathfrak{A}^*$ with* $\text{card}(A^*) \leq 2^\kappa$.

*Proof*: We shall construct an elementary $\kappa^+$-chain $(\mathfrak{A}_\nu)_{\nu<\kappa^+}$ with $\mathfrak{A}_0 = \mathfrak{A}$ and such that, for all $\nu < \kappa^+$:

(1) $\text{card}(A_\nu) \leq 2^\kappa$, and
(2) $\mathfrak{A}_{\nu+1}$ realizes all types $\Phi(v_0)$ of $\mathfrak{A}_\nu$ with $\text{card}(\Phi) \leq \kappa$.

Let $\mathfrak{A}^* = \mathfrak{A}_{\kappa^+}$ be the union of this chain. By Theorem 2.4.6, $\mathfrak{A}^*$ will be an elementary extension of $\mathfrak{A}$. From (1) will follow, in addition:

$$\text{card}(A^*) = \text{card}\left(\bigcup_{\nu<\kappa^+} A_\nu\right) \leq \max\{\kappa^+, 2^\kappa\} = 2^\kappa.$$

Finally, it will follow that $\mathfrak{A}^*$ is $\kappa^+$-saturated. Namely, if $\Phi(v_0)$ is a type of $(\mathfrak{A}^*,A')$ with $\text{card}(A') \leq \kappa$, then $A' \subseteq A_\nu$ must hold for some $\nu < \kappa^+$, by the regularity of the successor cardinal $\kappa^+$. From the fact that $\mathfrak{A}_\nu \preceq \mathfrak{A}^*$ will follow, then, that $\Phi(v_0)$ is also a type of $\mathfrak{A}_\nu$. In view of

$$\text{card}(\Phi) \leq \max\{\kappa_L, \text{card}(A')\} \leq \kappa,$$

we may apply (2) to conclude that an element $a \in A_{\nu+1}$ will realize this type in $\mathfrak{A}_{\nu+1}$, i.e. $\Phi(\underline{a})$ holds in $(\mathfrak{A}_{\nu+1}, A_{\nu+1})$. This will imply, however, using $\mathfrak{A}_{\nu+1} \preceq \mathfrak{A}^*$, that $(\mathfrak{A}^*, A^*) \models \Phi(\underline{a})$, as desired.

Thus, it remains only to construct the elementary $\kappa^+$-chain $(\mathfrak{A}_\nu)_{\nu<\kappa^+}$ satisfying (1) and (2). We begin with $\mathfrak{A}_0 = \mathfrak{A}$. In the case of a limit ordinal, we set $\mathfrak{A}_\lambda =$

$\bigcup_{\nu<\lambda} \mathfrak{A}_\nu$. In the case of a successor ordinal, $\mathfrak{A}_{\nu+1}$ will itself be obtained as the union of an elementary chain. For this, we index all types $\Phi(v_0)$ of $\mathfrak{A}_\nu$ with $\mathrm{card}(\Phi) \leq \kappa$ by means of ordinal numbers $\mu < 2^\kappa$; we obtain $(\Phi_\mu)_{\mu<2^\kappa}$. This is possible, since, first, $\mathrm{card}(\mathrm{Fml}(L(A_\nu))) \leq 2^\kappa$, and, second, the set of subsets of $\mathrm{Fml}(L(A_\nu))$ of cardinality $\leq \kappa$ can be bounded from above by the cardinality of the set of mappings from $\kappa$ into $\mathrm{Fml}(L(a_\nu))$. These two inequalities give:

$$\mathrm{card}(\mathrm{Fml}(L(A_\nu))^\kappa) \leq (2^\kappa)^\kappa = 2^{\kappa\cdot\kappa} = 2^\kappa,$$

as claimed. Now we set

$$\mathfrak{A}_{\nu+1} = \bigcup_{\mu<2^\kappa} \mathfrak{A}_\nu^{(\mu)},$$

where we define the sequence $\mathfrak{A}_\nu^{(\mu)}$ as follows:

$$\begin{aligned}
\mathfrak{A}_\nu^{(0)} &= \mathfrak{A}_\nu. \\
\mathfrak{A}_\nu^{(\lambda)} &= \bigcup_{\mu<\lambda} \mathfrak{A}_\nu^{(\mu)} \quad \text{for limit ordinals } \lambda < 2^\kappa. \\
\mathfrak{A}_\nu^{(\mu+1)} &= \text{an elementary extension } \mathfrak{B} \text{ of } \mathfrak{A}_\nu^{(\mu)} \\
&\quad\ \text{that realizes } \Phi_\mu(v_0), \text{ with } \mathrm{card}(B) \leq 2^\kappa.
\end{aligned} \tag{2.5.2.1}$$

The existence of an elementary extension $\mathfrak{B}$ as in (2.5.2.1) follows immediately from the definition of a type of $\mathfrak{A}_\nu^{(\mu)}$, together with Theorem 2.3.3.

By Theorem 2.4.6, $\mathfrak{A}_{\nu+1}$ is now an elementary extension of $\mathfrak{A}_\nu$. Condition (2) is obviously fulfilled, since for a type $\Phi(v_0)$ of $\mathfrak{A}_\nu$ with $\mathrm{card}(\Phi) \leq \kappa$, there is a $\mu < 2^\kappa$ with $\Phi = \Phi_\mu$. By construction (2.5.2.1) there is, then, an $a \in A_\nu^{(\mu+1)}$ that satisfies $\Phi$ in $A_\nu^{(\mu+1)}$. But then $\Phi(\underline{a})$ holds also in $(\mathfrak{A}_{\nu+1}, A_{\nu+1})$, since $\mathfrak{A}_\nu^{(\mu+1)} \preceq \mathfrak{A}_{\nu+1}$. Condition (1) follows from

$$\mathrm{card}(A_{\nu+1}) = \mathrm{card}\left(\bigcup_{\mu<2^\kappa} A_\nu^{(\mu)}\right) \leq \max\{2^\kappa, 2^\kappa\} = 2^\kappa. \qquad \square$$

*Remark 2.5.3* (Existence of "same-cardinality"-saturated, elementary extensions under GCH). We call an infinite $L$-structure $\mathfrak{A}$ *saturated* if it is $\mathrm{card}(|\mathfrak{A}|)$-saturated. If we assume the Generalized Continuum Hypothesis (GCH), $2^\kappa = \kappa^+$, for infinite cardinal numbers $\kappa$, then Theorem 2.5.2 says that to every structure $\mathfrak{A}$ and to every successor cardinal $\kappa^+$ with

$$\max\{\kappa_L^+, \mathrm{card}(|\mathfrak{A}|)\} \leq \kappa^+,$$

there is a saturated, elementary extension of cardinality $\kappa^+$. This structure is uniquely determined up to isomorphism, as Theorem 2.5.7 will show.

In order to make the proof of the isomorphism theorem easier to understand, we would like to insert the embedding theorem before it.

**Theorem 2.5.4** (Embedding Theorem). *Let $\mathfrak{A}$ and $\mathfrak{A}'$ be L-structures. Let $\mathfrak{A}'$ be $\kappa$-saturated, where $\kappa$ is an infinite cardinal $\geq \text{card}(|\mathfrak{A}|)$. Then if every $\exists$-sentence that holds in $\mathfrak{A}$ holds also in $\mathfrak{A}'$, then $\mathfrak{A}$ can be embedded in $\mathfrak{A}'$.*

Here, by an *existential* or *$\exists$-sentence* we mean an $L$-sentence of the form $\exists x_1,\ldots,x_n\,\delta$, where $\delta$ is quantifier-free.[3] The notation

$$\mathfrak{A} \overset{\exists}{\leadsto} \mathfrak{A}'$$

will denote the theorem's hypothesis that for every $\exists$-sentence, its truth in $\mathfrak{A}$ transfers to $\mathfrak{A}'$. We use this notation analogously in extension languages of $L$.

*Proof*: Let $A = |\mathfrak{A}|$, $A' = |\mathfrak{A}'|$, and $\alpha = \text{card}(A)$. We well-order the elements of $A$ by the transfinite sequence $(a_\nu)_{\nu<\alpha}$. We shall construct an $\alpha$-sequence $(a'_\nu)_{\nu<\alpha}$ of elements of $A'$ such that

$$(\mathfrak{A},(a_\nu)_{\nu<\alpha}) \overset{\exists}{\leadsto} (\mathfrak{A}',(a'_\nu)_{\nu<\alpha}). \qquad (2.5.4.1)_\alpha$$

The language $L_\alpha$ of this extension is just the extension of $L$ by constant symbols $c_\nu$ for $\nu < \alpha$, where we assume that $K \cap \alpha = K \cap \{\nu \mid \nu < \alpha\} = \emptyset$ (1.2.0.3). In the course of the proof we shall consider, analogously, the constant extensions $L_\beta$ of $L$ for all $\beta \leq \alpha$.

The embeddability of $\mathfrak{A}$ in $\mathfrak{A}'$ will now follow from $(2.5.4.1)_\alpha$, by Lemma 2.4.2. Indeed, $(\mathfrak{A}',(a'_\nu)_{\nu<\alpha})$ is a model of $D(\mathfrak{A})$, by $(2.5.4.1)_\alpha$. One need only be careful to notice that this time our constant symbols are not indexed by $a \in A$, but rather by the ordinal index $\nu$ of $a = a_\nu$.

We shall construct the sequence $(a'_\nu)_{\nu<\alpha}$ so that for all $\beta < \alpha$,

$$(\mathfrak{A},(a_\nu)_{\nu\leq\beta}) \overset{\exists}{\leadsto} (\mathfrak{A}',(a'_\nu)_{\nu\leq\beta}). \qquad (2.5.4.2)_\beta$$

This implies $(2.5.4.1)_\alpha$ immediately, as follows. If $\alpha$ is a finite cardinal number, this is immediately clear (take $\beta = \alpha - 1$). If $\alpha$ is an infinite cardinal number, then $\alpha$ is a limit ordinal. Therefore, since only finitely many constant symbols $c_\nu$ can occur in any one $\exists$-formula $\varphi$ in the language $L_\alpha$, $\varphi$ is already an $\exists$-formula in the language $L_\beta$ for some $\beta < \alpha$. Therefore, from $(2.5.4.2)_\beta$ and the hypothesis $(\mathfrak{A},(a_\nu)_{\nu<\alpha}) \models \varphi$ we immediately obtain $(\mathfrak{A}',(a'_\nu)_{\nu<\alpha}) \models \varphi$.

Now to the definition of the sequence $(a'_\nu)_{\nu<\alpha}$. Let $\gamma < \alpha$. We assume that the sequence $(a'_\nu)_{\nu<\gamma}$ has already been defined so that for all $\beta < \gamma$, $(2.5.4.2)_\beta$ holds. Then for $a_\gamma$ we must find an $a'_\gamma \in A'$ such that $(2.5.4.2)_\gamma$ holds.

Let $\Phi(v_0)$ be the set of all $\exists$-formulae $\varphi$ of the language $L_\gamma$ with $\text{Fr}(\varphi) \subseteq \{v_0\}$ and $(\mathfrak{A},(a_\nu)_{\nu\leq\gamma}) \models \varphi(c_\gamma)$. Obviously, $\Phi(v_0)$ is a type of $(\mathfrak{A},(a_\nu)_{\nu<\gamma})$. So for each finite subset $\{\varphi_1,\ldots,\varphi_n\}$ of $\Phi(v_0)$ we have

---

[3] The case $n = 0$ thus includes quantifier-free sentences. Note also that this concept differs from the "existence sentences" introduced in Theorem 1.4.2; the latter require $n > 0$ and do not require $\delta$ to be quantifier-free.

$$(\mathfrak{A},(a_\nu)_{\nu<\gamma}) \models \exists v_0\,(\varphi_1 \wedge \cdots \wedge \varphi_n)$$

(2.5.1). Let $\beta$ be the maximum of the indices of the constants $c_\nu$ occurring in $\exists v_0\,(\varphi_1 \wedge \cdots \wedge \varphi_n)$. Obviously, $\beta < \gamma$, and hence $\exists v_0\,(\varphi_1 \wedge \cdots \wedge \varphi_n)$ is logically equivalent to an $\exists$-sentence in the language $L_{\beta+1}$. By hypothesis we also thereby conclude

$$(\mathfrak{A}',(a'_\nu)_{\nu<\gamma}) \models \exists v_0\,(\varphi_1 \wedge \cdots \wedge \varphi_n).$$

Therefore $\Phi(v_0)$ is also a type of $(\mathfrak{A}',(a'_\nu)_{\nu<\gamma})$, again using (2.5.1). By the $\kappa$-saturation of $\mathfrak{A}'$ we obtain the $\kappa$-saturation of $(\mathfrak{A}',(a'_\nu)_{\nu<\gamma})$, as already mentioned just after the definition of $\kappa$-saturation (p. 82). Since $\mathrm{card}(\gamma) < \alpha \le \kappa$ (2.5.1.3), $\Phi(v_0)$ is realized in this structure, i.e. there is an $a'_\gamma \in A'$ with $(\mathfrak{A}',(a'_\nu)_{\nu\le\gamma}) \models \Phi(c_\gamma)$. But this shows $(2.5.4.2)_\gamma$. Indeed, writing an $\exists$-sentence $\varphi$ in the form $\exists x_1,\ldots,x_n\,\delta$ with $\delta$ quantifier-free, we can always, without loss of generality, assume $v_0 \notin \{x_1,\ldots,x_n\}$ and thereby obtain $\varphi = \varphi(c_\gamma/v_0)(v_0/c_\gamma)$. □

Before we come to the Isomorphism Theorem, we would still like to deduce several consequences of the above theorem. Here we call an $L$-substructure $\mathfrak{A}$ of $\mathfrak{A}'$ *existentially closed* if every $\exists$-sentence of the language $L(\mathfrak{A})$ that holds in $\mathfrak{A}'$ also holds in $\mathfrak{A}$.

**Corollary 2.5.5.** *Let the L-structure $\mathfrak{A}$ with universe A be a common substructure of the L-structures $\mathfrak{A}^*$ and $\mathfrak{A}'$.*

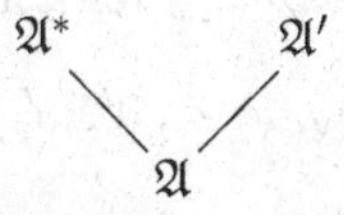


(1) *If $\mathfrak{A}'$ is $\kappa$-saturated with $\kappa > \mathrm{card}(|\mathfrak{A}^*|)$, and either*
   (a) *$\mathfrak{A}$ is existentially closed in $\mathfrak{A}^*$ or*
   (b) *every finitely generated $L(A)$-substructure of $\mathfrak{A}^*$ is embeddable in $\mathfrak{A}'$,*
   *then $\mathfrak{A}^*$ is embeddable in $\mathfrak{A}'$ as an $L(A)$-structure.*

(2) *If $\mathfrak{A}^*$ is embeddable in $\mathfrak{A}'$ as an $L(A)$-structure, and $\mathfrak{A} \preceq \mathfrak{A}'$, then $\mathfrak{A}$ is existentially closed in $\mathfrak{A}^*$.*

*Proof*: (1) By Theorem 2.5.4, it suffices to show $(\mathfrak{A}^*,A) \overset{\exists}{\rightsquigarrow} (\mathfrak{A}',A)$. Observe that the structure $(\mathfrak{A}',A)$ is also $\kappa$-saturated, since $\mathrm{card}(A) \le \mathrm{card}(A^*) < \kappa$ (where we write $A^* = |\mathfrak{A}^*|$). Now let $\varphi$ be an $\exists$-sentence of $L(A)$ with $(\mathfrak{A}^*,A) \models \varphi$. If $\mathfrak{A}$ is existentially closed in $\mathfrak{A}^*$, then $(\mathfrak{A},A) \models \varphi$, whence $(\mathfrak{A}',A) \models \varphi$ by the existential character of $\varphi$. Thus we are done in this case.

Now assume the embeddability of finitely generated $L(A)$-substructures of $\mathfrak{A}^*$ in $\mathfrak{A}'$. Writing $\varphi$ in the form $\exists x_1,\ldots,x_n\,\delta$ with $\delta$ quantifier-free, from $(\mathfrak{A}^*,A) \models \varphi$ follows the existence of elements $a_1^*,\ldots,a_n^* \in A^*$ with $(\mathfrak{A}^*,A^*) \models \delta(a_1^*,\ldots,a_n^*)$. Let $\mathfrak{B}$ be the substructure of $\mathfrak{A}^*$ generated by $A \cup \{a_1^*,\ldots,a_n^*\}$. Then $\mathfrak{A} \subseteq \mathfrak{B} \subseteq \mathfrak{A}^*$, and $\mathfrak{B}$ is a finitely generated $L(A)$-substructure of $\mathfrak{A}^*$. By hypothesis there is an embedding $\tau : \mathfrak{B} \to \mathfrak{A}'$ with $\tau(a) = a$ for all $a \in A$. Then according to Remark 2.2.2 and Lemma 2.4.1, $(\mathfrak{A}',A) \models \delta(\tau(a_1^*),\ldots,\tau(a_n^*))$, whence, in particular,

$$(\mathfrak{A}', A) \models \exists x_1, \ldots, x_n \, \delta. \tag{2.5.5.1}$$

(2) If $\varphi$ is an existential $L(A)$-sentence with $(\mathfrak{A}^*, A) \models \varphi$, then from the $L(A)$-embeddability of $\mathfrak{A}^*$ in $\mathfrak{A}'$ follows naturally $(\mathfrak{A}', A) \models \varphi$, by analogy with (2.5.5.1). This, however, gives $(\mathfrak{A}, A) \models \varphi$, using $\mathfrak{A} \preceq \mathfrak{A}'$. □

The next corollary follows immediately from the proof of the Embedding Theorem 2.5.4, along with Lemma 2.4.2, when one takes for $\Phi(v_0)$ not the set of all $\exists$-formulae, but rather all formulae $\varphi$ of the language $L_\gamma$ with $\mathrm{Fr}(\varphi) \subseteq \{v_0\}$ and $(\mathfrak{A}, (a_\nu)_{\nu<\gamma}) \models \varphi(c_\gamma)$.

**Corollary 2.5.6.** *Let $\mathfrak{A}$ and $\mathfrak{A}'$ be L-structures. Let $\mathfrak{A}'$ be $\kappa$-saturated, where $\kappa$ is an infinite cardinal $\geq \mathrm{card}(|\mathfrak{A}|)$. Then if $\mathfrak{A}$ and $\mathfrak{A}'$ are elementarily equivalent, then $\mathfrak{A}$ can be elementarily embedded in $\mathfrak{A}'$.*

Now we come to the important:

**Theorem 2.5.7** (Isomorphism Theorem). *Let $\mathfrak{A}$ and $\mathfrak{A}'$ be infinite, saturated L-structures of equal cardinality $\kappa$. Then $\mathfrak{A}$ and $\mathfrak{A}'$ are isomorphic if and only if they are elementarily equivalent.*

*Proof*: If $\mathfrak{A}$ and $\mathfrak{A}'$ are isomorphic, then it follows immediately that they are elementary equivalent, by Theorem 2.2.1. So suppose, now, that $\mathfrak{A}$ and $\mathfrak{A}'$ are elementarily equivalent. We shall well-order the elements of $A := |\mathfrak{A}|$ and $A' := |\mathfrak{A}'|$ by the transfinite sequences $(a_\nu)_{\nu<\kappa}$ and $(a'_\nu)_{\nu<\kappa}$, respectively, in such a way that

$$(\mathfrak{A}, (a_\nu)_{\nu<\kappa}) \equiv (\mathfrak{A}', (a'_\nu)_{\nu<\kappa}) \tag*{(2.5.7.1)$_\kappa$}$$

holds in the language $L_\kappa$ with constants $c_\nu$ for $\nu < \kappa$. (Without loss of generality, we assume, further, that $K \cap \kappa = K \cap \{\nu \mid \nu < \kappa\} = \emptyset$ (1.2.0.3).) Then by Lemma 2.4.2, the mapping $\tau(a_\nu) = a'_\nu$ will define an embedding of $\mathfrak{A}$ into $\mathfrak{A}'$, which is obviously surjective, i.e. $\tau$ is an isomorphism of $\mathfrak{A}$ onto $\mathfrak{A}'$. This follows as in the proof of Theorem 2.5.4, since $\mathfrak{A}$ is a model of $D(\mathfrak{A})$, by (2.5.7.1)$_\kappa$. We shall therefore alter the proof of the Embedding Theorem 2.5.4 in such a way that (2.5.7.1)$_\kappa$ holds.

We begin with well-orderings $(b_\nu)_{\nu<\kappa}$ and $(b'_\nu)_{\nu<\kappa}$ of $A$ and $A'$, respectively. From these well-orderings we shall construct the well-orderings $(a_\nu)_{\nu<\kappa}$ and $(a'_\nu)_{\nu<\kappa}$ so that

$$(\mathfrak{A}, (a_\nu)_{\nu\leq\beta}) \equiv (\mathfrak{A}', (a'_\nu)_{\nu\leq\beta}) \tag*{(2.5.7.2)$_\beta$}$$

holds for all $\beta < \kappa$. From this (2.5.7.1)$_\kappa$ will follow immediately: indeed, since $\kappa$, being a cardinal number, is a limit ordinal, and every $L_\kappa$-sentence $\varphi$ can contain only finitely many constant symbols $c_\nu$ with $\nu < \kappa$, $\varphi$ is already an $L_\beta$-sentence for some $\beta < \kappa$.

We shall define the transfinite sequences $(a_\nu)_{\nu<\kappa}$ and $(a'_\nu)_{\nu<\kappa}$ so that eventually every $b_\nu$ and every $b'_\nu$ occurs in these sequences. In order to achieve this, we shall switch back and forth between the structures $\mathfrak{A}$ and $\mathfrak{A}'$ (compare Theorem 3.2.3(1) below). Here we shall use the fact that every ordinal number $\gamma$ can be represented uniquely in the form

$$\gamma = \lambda + m, \tag{2.5.7.3}$$

where $\lambda$ is either 0 or a limit ordinal, and $m \in \mathbb{N}$.

Now suppose, using definition by recursion, that $\gamma < \kappa$ and the sequences $(a_\nu)$ and $(a'_\nu)$ have already been defined for $\nu < \gamma$ so that $(2.5.7.2)_\beta$ holds for all $\beta < \gamma$. We shall define $a_\gamma$ and $a'_\gamma$ so that $(2.5.7.2)_\gamma$ holds. Write the ordinal $\gamma$ as in (2.5.7.3).

*Case 1*: $m$ is even; let us write $m = 2n$.

In this case let $a_\gamma$ be that element $b_\nu$ of the set

$$\{b_\nu \mid \nu < \kappa\} \setminus \{a_\nu \mid \nu < \gamma\}$$

with smallest index $\nu$. In order to define $a'_\gamma$, we consider the type of $a_\gamma$ in $(\mathfrak{A}, (a_\nu)_{\nu \leq \gamma})$, i.e. the set $\Phi(v_0)$ of formulae of the language $L_\gamma$ for which

$$(\mathfrak{A}, (a_\nu)_{\nu \leq \gamma}) \models \Phi(c_\gamma)$$

holds. Thus if $\varphi_1, \ldots, \varphi_r \in \Phi(v_0)$, then $(\mathfrak{A}, (a_\nu)_{\nu < \gamma}) \models \exists v_0 (\varphi_1 \wedge \cdots \wedge \varphi_r)$. Let $\beta$ be the maximum of the indices of the constant symbols $c_\nu$ in $(\varphi_1 \wedge \cdots \wedge \varphi_r)$. Since $\beta < \gamma$, $(2.5.7.2)_\beta$ holds, and thus $(\mathfrak{A}', (a'_\nu)_{\nu < \gamma}) \models \exists v_0 (\varphi_1 \wedge \cdots \wedge \varphi_r)$. Therefore $\Phi(v_0)$ is a type of $(\mathfrak{A}, (a_\nu)_{\nu < \gamma})$ (2.5.1). From the $\kappa$-saturation of $\mathfrak{A}'$ and $\mathrm{card}(\gamma) < \kappa$ follows, then, that there is a $b'_\mu \in A'$ that realizes $\Phi(v_0)$ in $(\mathfrak{A}', (a'_\nu)_{\nu < \gamma})$. Let $a'_\gamma$ be such a $b'_\mu$ with smallest index $\mu$. Then obviously

$$(\mathfrak{A}', (a'_\nu)_{\nu \leq \gamma}) \models \Phi(c_\gamma). \tag{2.5.7.4}$$

From this, $(2.5.7.2)_\gamma$ follows. Indeed, if $\varphi$ is an $L_{\gamma+1}$-sentence, then one easily finds, by possibly renaming bound variables, an $L_{\gamma+1}$-sentence $\varphi'$ such that $\varphi$ and $\varphi'$ are logically equivalent and $c_\gamma$ does not lie anywhere inside the scope of a quantifier $\exists v_0$. Therefore, we may assume, without loss of generality, that $\varphi$ equals $\varphi(c_\gamma/v_0)(v_0/c_\gamma)$. Then if $(\mathfrak{A}, (a_\nu)_{\nu \leq \gamma}) \models \varphi$, then $\varphi(c_\gamma/v_0) \in \Phi(v_0)$, and this implies $(\mathfrak{A}', (a'_\nu)_{\nu \leq \gamma}) \models \varphi$, using (2.5.7.4).

*Case 2*: $m$ is odd; let us write $m = 2n + 1$.

In this case let $a'_\gamma$ be that element $b'_\nu$ of the set

$$\{b'_\nu \mid \nu < \kappa\} \setminus \{a'_\nu \mid \nu < \gamma\}$$

with smallest index $\nu$. We now define $a_\gamma$ completely analogously to our definition of $a'_\gamma$ in Case 1, and we again obtain

$$(\mathfrak{A}, (a_\nu)_{\nu \leq \gamma}) \equiv (\mathfrak{A}', (a'_\nu)_{\nu \leq \gamma}).$$

Now one sees immediately, from the alternating character of the definitions of the sequences $(a_\nu)_{\nu < \kappa}$ and $(a'_\nu)_{\nu < \kappa}$, that in Case 1, the smallest index $\nu$ with $b_\nu \notin \{a_\mu \mid \mu < \gamma\}$ must obviously satisfy $\nu \geq \lambda + n$. In the second case, analogously, the smallest index $\nu$ with $b'_\nu \notin \{a'_\mu \mid \mu < \gamma\}$ satisfies $\nu \geq \lambda + n$. Therefore all $b_\nu$ and $b'_\nu$ will, eventually, be chosen. □

To close out this section, we would like to consider several examples of saturated structures.

Let $\mathfrak{A} = \langle A; <^{\mathfrak{A}} \rangle$ be a dense linear ordering without extrema, i.e. a model of the axioms $O_1$ through $O_5$ in Section 1.6(1). For $<^{\mathfrak{A}}$ we shall simply write $<$ here. Let $\mathfrak{A}$ be $\kappa$-saturated, where $\kappa = \aleph_\alpha$. In this case, we claim that $\mathfrak{A}$ has "order-type $\eta_\alpha$", a property introduced by Hausdorff; namely, if $B$ and $C$ are subsets of $A$ with $B < C$ (this means that $b < c$ for all $b \in B$ and $c \in C$) and cardinalities $< \aleph_\alpha$, then there exists an $a \in A$ with

$$b < a < c \text{ for all } b \in B,\ c \in C.$$

To see this, we consider the set

$$\Phi := \Phi(v_0) = \{\underline{b} < v_0 \mid b \in B\} \cup \{v_0 < \underline{c} \mid c \in C\}$$

of formulae. Since $\mathfrak{A} = \langle A; < \rangle$ is a dense linear ordering without extrema, every finite subset of $\Phi$ can obviously be realized in $\mathfrak{A}$. Thus, since $\text{card}(\Phi) < \aleph_\alpha = \kappa$, $\Phi$ itself can be realized in $\mathfrak{A}$. Then a realizing element $a \in A$ satisfies exactly $b < a < c$ for every $b \in B$ and $c \in C$. (Let us point out that both $B$ and $C$ may be empty.)

The order-type $\eta_0$ is clearly just the type of dense orderings without extrema. The order-type $\eta_1$, however, cannot be illustrated so easily with examples: the orderings of $\mathbb{Q}$ and $\mathbb{R}$, respectively, are only of type $\eta_0$.

In Chapter 4 we shall see that a linear ordering $\mathfrak{A} = \langle A; < \rangle$ of order-type $\eta_\alpha$ is always $\aleph_\alpha$-saturated. We shall establish the analogous result for divisible, ordered Abelian groups and for real closed fields. If, say, $\mathfrak{A} = \langle A; <^{\mathfrak{A}}; +^{\mathfrak{A}}; 0^{\mathfrak{A}} \rangle$ is a divisible, ordered Abelian group (recall Section 1.6), then the $\aleph_\alpha$-saturation of $\mathfrak{A}$ will imply the order-type $\eta_\alpha$ of $\langle A; <^{\mathfrak{A}} \rangle$, as above. As we shall see, order-type $\eta_\alpha$ is, on the other hand, also sufficient for the $\aleph_\alpha$-saturation of $\mathfrak{A}$. Likewise we shall see that a real closed field $\mathfrak{A} = \langle A; <^{\mathfrak{A}}; +^{\mathfrak{A}}, \cdot^{\mathfrak{A}}; 0^{\mathfrak{A}}, 1^{\mathfrak{A}} \rangle$ (recall Section 1.6(6)) is $\aleph_\alpha$-saturated if and only if $\langle A; <^{\mathfrak{A}} \rangle$ has order-type $\eta_\alpha$.

The statements just made are (as the reader will easily see later) direct consequences of the fact that the theories referred to admit "elimination of quantifiers" (see Section 3.4).

## 2.6 Ultraproducts

In this section we shall introduce a very algebraically appealing construction of an $L$-structure from one or more given $L$-structures. This method will allow us to, among other things, "weld" models of the finite subsets of a set $\Sigma$ of $L$-sentences together into a model of $\Sigma$ itself. By means of this we obtain a proof of the Finiteness Theorem 1.5.6 that does not rely upon the Completeness Theorem.

As preparation we must introduce the concept of an ultrafilter on a set. Let $S$ be a nonempty set, and $P(S)$ the power set of $S$, i.e. $P(S) = \{A \mid A \subseteq S\}$. A nonempty subset $\mathscr{F}$ of $P(S)$ is called a *filter* on $S$ if

$$\begin{aligned} &(1)\quad \emptyset \notin \mathscr{F}, \\ &(2)\quad U,V \in \mathscr{F} \text{ implies } U \cap V \in \mathscr{F}, \text{ and} \\ &(3)\quad U \in \mathscr{F},\ U \subseteq A \subseteq S \text{ implies } A \in \mathscr{F}. \end{aligned} \tag{2.6.0.1}$$

Examples of filters include the set of all co-finite subsets $A$ of $S$ (i.e. those $A$ such that $S \setminus A$ is finite; here $S$ is assumed to be infinite), or, for each $a \in S$, the set

$$\mathscr{H}(a) = \{A \subseteq S \mid a \in A\}. \tag{2.6.0.2}$$

The filter $\mathscr{F} = \mathscr{H}(a)$ has the additional property

$$(4)\quad A \subseteq S,\ A \notin \mathscr{F} \text{ implies } S \setminus A \in \mathscr{F}. \tag{2.6.0.3}$$

A filter on $S$ satisfying (4) is called an *ultrafilter* on $S$. Filters of the form $\mathscr{H}(a)$ are called *principal* ultrafilters. One can see easily that an ultrafilter $\mathscr{F}$ that contains (as an element) a finite set, is necessarily principal. If $S$ is infinite, then the filter of co-finite subsets of $S$ is obviously not an ultrafilter (cf. Exercise 2.7.8). However, the following lemma holds:

**Lemma 2.6.1.** *Let $\mathscr{F}_0$ be a system of nonempty subsets of $S$ that is closed under (finite) intersection. Then there is an ultrafilter $\mathscr{D}$ on $S$ with $\mathscr{F}_0 \subseteq \mathscr{D}$.*

*Proof*: Let $\mathscr{D} \subseteq P(S) \setminus \{\emptyset\}$ be a maximal, intersection-closed extension of $\mathscr{F}_0$. Zorn's lemma guarantees the existence of such a system $\mathscr{D}$. We shall show that $\mathscr{D}$ is an ultrafilter on $S$.

If we form the set

$$\mathscr{F}(\mathscr{D}) = \{A \subseteq S \mid \text{there exists a } U \in \mathscr{D} \text{ with } U \subseteq A\}, \tag{2.6.1.1}$$

then we see immediately that $\mathscr{F}(\mathscr{D})$ is an intersection-closed subset of $P(S) \setminus \{\emptyset\}$ extending $\mathscr{D}$. Therefore

$$\mathscr{F}(\mathscr{D}) = \mathscr{D}, \tag{2.6.1.2}$$

by the maximality of $\mathscr{D}$. And since $\mathscr{F}(\mathscr{D})$ obviously satisfies (3), we conclude that $\mathscr{D}$ is a filter on $S$. Now suppose $A \subseteq S$ and $A \notin \mathscr{D}$. Set $B = S \setminus A$ and form the set

$$\mathscr{F}[B] := \mathscr{D} \cup \{U \cap B \mid U \in \mathscr{D}\},$$

which is obviously intersection-closed. If $\emptyset$ were a member of $\mathscr{F}[B]$, then $U \cap B$ would be empty for some $U \in \mathscr{D}$. Then we would have $U \subseteq A = S \setminus B$, which, in view of (2.6.1.1) and (2.6.1.2), would contradict $A \notin \mathscr{D}$. So we may invoke the maximality of $\mathscr{D}$ to conclude that $\mathscr{F}[B]$ must again be equal to $\mathscr{D}$. Since $S \in \mathscr{F}(\mathscr{D}) = \mathscr{D}$, we conclude that

$$S \setminus A = B = S \cap B \in \mathscr{F}[B] = \mathscr{D}.$$

Thus we have proved that $\mathscr{D}$ is an ultrafilter. □

Now let a nonempty system $\mathfrak{A}^{(s)}$ of $L$-structures, for $s \in S$, be given. Further let an ultrafilter $\mathscr{D}$ on $S$ be given. Then we shall define the *ultraproduct of the* $\mathfrak{A}^{(s)}$ *with respect to* $\mathscr{D}$. This will again be an $L$-structure. For this, let

$$\mathfrak{A}^{(s)} = \langle \mathfrak{A}^{(s)}; (\mathscr{R}_i^{(s)})_{i\in I}; (\mathsf{f}_j^{(s)})_{j\in J}; (d_k^{(s)})_{k\in K} \rangle.$$

We consider the following set of sequences indexed by $s \in S$:

$$\prod_{s\in S} A^{(S)} = \{ (a^{(s)})_{s\in S} \mid a^{(s)} \in \mathfrak{A}^{(s)} \text{ for all } s \in S \}.$$

We would like to call two sequences $(a_1^{(s)})_{s\in S}$ and $(a_2^{(s)})_{s\in S}$ *equivalent* (with respect to $\mathscr{D}$) if they agree with each other on a subset from $\mathscr{D}$; i.e. we set

$$(a_1^{(s)})_{s\in S} \underset{\mathscr{D}}{\sim} (a_2^{(s)})_{s\in S} \quad \text{iff} \quad \{ s \in S \mid a_1^{(s)} = a_2^{(s)} \} \in \mathscr{D}.$$

(Usually we write $(a^{(s)})$ instead of $(a^{(s)})_{s\in S}$, and we write $\sim$ instead of $\underset{\mathscr{D}}{\sim}$.) The relation $\sim$ is an equivalence relation on $\prod_{s\in S} \mathfrak{A}^{(S)}$; i.e.

(i) $(a_1^{(s)}) \sim (a_1^{(s)})$;

(ii) $(a_1^{(s)}) \sim (a_2^{(s)})$ implies $(a_2^{(s)}) \sim (a_1^{(s)})$; and

(iii) $(a_1^{(s)}) \sim (a_2^{(s)})$ and $(a_2^{(s)}) \sim (a_3^{(s)})$ imply $(a_1^{(s)}) \sim (a_3^{(s)})$.

Here (i) follows from $S \in \mathscr{D}$, (ii) is trivial, and (iii) results as follows:

$$\{ s \in S \mid a_1^{(s)} = a_2^{(s)} \} \cap \{ s \in S \mid a_2^{(s)} = a_3^{(s)} \} \subseteq \{ s \in S \mid a_1^{(s)} = a_3^{(s)} \};$$

since the two sets of the left-hand side lie in $\mathscr{D}$, so does their intersection, and thus also the set on the right-hand side.

Now we define the universe $A$ to be the set of equivalence classes of $\prod_{s\in S} \mathfrak{A}^{(S)}$ with respect to $\underset{\mathscr{D}}{\sim}$; i.e.

$$A = \left\{ \overline{(a^{(s)})} \;\middle|\; (a^{(s)}) \in \prod_{s\in S} A^{(S)} \right\},$$

where

$$\overline{(a^{(s)})} = \{ (a_1^{(s)}) \mid (a_1^{(s)}) \sim (a^{(s)}) \}. \tag{2.6.1.3}$$

On $A$ we shall define the relations $\mathscr{R}_i$, the functions $\mathsf{f}_j$, and the constants $d_k$, for $i \in I$, $j \in J$, and $k \in K$, respectively, so that we can construct the $L$-structure

$$\mathfrak{A} = \langle A; (\mathscr{R}_i)_{i\in I}; (\mathsf{f}_j)_{j\in J}; (d_k)_{k\in K} \rangle.$$

$\mathfrak{A}$ is called the *ultraproduct of the* $\mathfrak{A}^{(s)}$ *with respect to the ultrafilter* $\mathscr{D}$. For this we also write

$$\mathfrak{A} = \prod_{s \in S} \mathfrak{A}^{(s)} \Big/ \mathscr{D}.$$

We shall define the relations $\mathscr{R}_i$ as follows: for equivalence classes $\overline{(a_1^{(s)})}, \ldots, \overline{(a_{\lambda(i)}^{(s)})}$ we set

$$\mathscr{R}_i\left(\overline{(a_1^{(s)})}, \ldots, \overline{(a_{\lambda(i)}^{(s)})}\right) \quad \text{iff} \quad \{ s \mid \mathscr{R}_i^{(s)}(a_1^{(s)}, \ldots, a_{\lambda(i)}^{(s)}) \} \in \mathscr{D}. \tag{2.6.1.4}$$

This definition makes use of representatives of the equivalence classes. We must therefore show that the relation does not change under a different choice of representatives. So let $(b_1^{(s)}) \sim (a_1^{(s)})$, ..., $(b_{\lambda(i)}^{(s)}) \sim (a_{\lambda(i)}^{(s)})$. Then

$$\{ s \mid b_\nu^{(s)} = a_\nu^{(s)} \} \in \mathscr{D} \quad \text{for all } \nu \in \{1, 2, \ldots, \lambda(i)\}.$$

From this we obtain

$$U := \bigcap_{\nu=1}^{\lambda(i)} \{ s \mid b_\nu^{(s)} = a_\nu^{(s)} \} \in \mathscr{D}.$$

Now if $\mathscr{R}_i\left(\overline{(a_1^{(s)})}, \ldots, \overline{(a_{\lambda(i)}^{(s)})}\right)$ holds, then we have $\{ s \mid \mathscr{R}_i^{(s)}(a_1^{(s)}, \ldots, a_{\lambda(i)}^{(s)}) \} \in \mathscr{D}$, and by

$$\{ s \mid \mathscr{R}_i^{(s)}(a_1^{(s)}, \ldots, a_{\lambda(i)}^{(s)}) \} \cap U \subseteq \{ s \mid \mathscr{R}_i^{(s)}(b_1^{(s)}, \ldots, b_{\lambda(i)}^{(s)}) \}$$

it follows that also $\{ s \mid \mathscr{R}_i^{(s)}(b_1^{(s)}, \ldots, b_{\lambda(i)}^{(s)}) \} \in \mathscr{D}$, i.e. $\mathscr{R}_i\left(\overline{(b_1^{(s)})}, \ldots, \overline{(b_{\lambda(i)}^{(s)})}\right)$. One shows the converse analogously.

We define the functions $\mathsf{f}_j$ on a $\mu(j)$-tuple $\left(\overline{(a_1^{(s)})}, \ldots, \overline{(a_{\mu(j)}^{(s)})}\right)$ of equivalence classes, via

$$\mathsf{f}_j\left(\overline{(a_1^{(s)})}, \ldots, \overline{(a_{\mu(j)}^{(s)})}\right) := \overline{(\mathsf{f}_j^{(s)}(a_1^{(s)}, \ldots, a_{\mu(j)}^{(s)}))}.$$

We proceed analogously with the constants $d_k$:

$$d_k := \overline{(d_k^{(s)})}.$$

In the case of the functions $\mathsf{f}_j$ for $j \in J$, one must again show that the definition is independent of the chosen representatives. We leave this to the reader.

The following theorem contains the most important property of ultraproducts. In its formulation we require the concept of a "sequence of evaluations". For $s \in S$, let $h^{(s)}$ be an evaluation of the variables in $\mathfrak{A}^{(s)}$. Then we define

$$\overline{(h^{(s)})} : \mathrm{Vbl} \to A$$

by $\overline{(h^{(s)})}(v_i) := \overline{(h^{(s)}(v_i))}$. If $a = \overline{(a^{(s)})}$ is an element of $A$, then

$$\overline{(h^{(s)})}\binom{x}{a} = \overline{(h^{(s)}\binom{x}{a^{(s)}})},$$

as one may easily check. Then we have the following theorem of Łos:

**Theorem 2.6.2** (Łos' Theorem). *Let $\mathfrak{A}$ be the ultraproduct of the L-structures $\mathfrak{A}^{(s)}$, for $s \in S$, with respect to the ultrafilter $\mathscr{D}$. Then for every L-formula $\varphi$ and every sequence of evaluations $\overline{(h^{(s)})}$ in $\mathfrak{A}$,*

$$\mathfrak{A} \models \varphi\left[\overline{(h^{(s)})}\right] \quad \textit{iff} \quad \{s \mid \mathfrak{A}^{(s)} \models \varphi[h^{(s)}]\} \in \mathscr{D}. \tag{2.6.2.1}$$

*Proof*: We prove (2.6.2.1) by induction on the recursive construction of $\varphi$.

First suppose $\varphi$ is atomic. If $\varphi$ is $\mathscr{R}_i(t_1,\ldots,t_{\lambda(i)})$, then (2.6.2.1) follows from (2.6.1.4), (1.5.0.6), and

$$t^{\mathfrak{A}}\left[\overline{(h^{(s)})}\right] = \overline{(t^{\mathfrak{A}^{(s)}}[h^{(s)}])},$$

which one proves by a routine induction on the recursive construction of terms (recalling (1.5.0.4)). When $\varphi$ is the other kind of atomic formula, $t_1 \doteq t_2$, (2.6.2.1) follows similarly.

If $\varphi$ is of the form $\neg\varphi_1$, then

$$\begin{aligned}
\mathfrak{A} \models \neg\varphi_1\left[\overline{(h^{(s)})}\right] \text{ iff } & \quad \mathfrak{A} \not\models \varphi_1\left[\overline{(h^{(s)})}\right] \\
\text{iff } & \quad \{s \in S \mid \mathfrak{A}^{(s)} \models \varphi_1[h^{(s)}]\} \notin \mathscr{D} \\
\text{iff } & S \setminus \{s \in S \mid \mathfrak{A}^{(s)} \models \varphi_1[h^{(s)}]\} \in \mathscr{D} \\
\text{iff } & \quad \{s \in S \mid \mathfrak{A}^{(s)} \not\models \varphi_1[h^{(s)}]\} \in \mathscr{D} \\
\text{iff } & \quad \{s \in S \mid \mathfrak{A}^{(s)} \models \neg\varphi_1[h^{(s)}]\} \in \mathscr{D}.
\end{aligned}$$

Here we used the induction hypothesis as well as the ultrafilter properties (4) and (1) ((2.6.0.3) and (2.6.0.1)).

If $\varphi$ is of the form $(\varphi_1 \wedge \varphi_2)$, then

$$\begin{aligned}
\mathfrak{A} \models \varphi\left[\overline{(h^{(s)})}\right] \quad & \text{iff } \left(\mathfrak{A} \models \varphi_1\left[\overline{(h^{(s)})}\right] \text{ and } \mathfrak{A} \models \varphi_2\left[\overline{(h^{(s)})}\right]\right) \\
& \text{iff } (\{s \mid \mathfrak{A}^{(s)} \models \varphi_1[h^{(s)}]\} \in \mathscr{D} \;\&\; \{s \mid \mathfrak{A}^{(s)} \models \varphi_2[h^{(s)}]\} \in \mathscr{D}) \\
& \text{iff } \{s \mid \mathfrak{A}^{(s)} \models \varphi_1[h^{(s)}]\} \cap \{s \mid \mathfrak{A}^{(s)} \models \varphi_2[h^{(s)}]\} \in \mathscr{D} \\
& \text{iff } \{s \mid \mathfrak{A}^{(s)} \models (\varphi_1 \wedge \varphi_2)[h^{(s)}]\} \in \mathscr{D}.
\end{aligned}$$

Here we have used the induction hypothesis as well as the filter properties (2) and (3) (2.6.0.1).

If $\varphi$ is of the form $\forall x\, \varphi_1$, then

$$\begin{aligned}
\mathfrak{A} \models \varphi\left[\overline{(h^{(s)})}\right] \quad &\text{iff} \quad \mathfrak{A} \models \varphi_1\left[\overline{(h^{(s)})}\binom{x}{a}\right] && \text{for all } a \in A\\
&\text{iff} \quad \mathfrak{A} \models \varphi_1\left[\overline{(h^{(s)})\binom{x}{a^{(s)}}}\right] && \text{for all } (a^{(s)}) \in \prod_{x\in S} A^{(s)}\\
&\text{iff } \left\{ s \mid \mathfrak{A}^{(s)} \models \varphi_1\left[h^{(s)}\binom{x}{a^{(s)}}\right]\right\} \in \mathscr{D} && \text{for all } (a^{(s)}) \in \prod_{x\in S} A^{(s)}\\
&\text{iff } \left\{ s \mid \mathfrak{A}^{(s)} \models \varphi_1\left[h^{(s)}\binom{x}{a^{(s)}}\right] \text{ for all } a^{(s)} \in A^{(s)} \right\} \in \mathscr{D}\\
&\text{iff } \left\{ s \mid \mathfrak{A}^{(s)} \models \forall x \varphi_1\left[h^{(s)}\right]\right\} \in \mathscr{D}.
\end{aligned}$$

The penultimate step above requires justification. For this we set

$$U = \left\{ s \mid \mathfrak{A}^{(s)} \models \varphi_1\left[h^{(s)}\binom{x}{a^{(s)}}\right] \text{ for all } a^{(s)} \in A^{(s)} \right\}.$$

Then for every sequence $(b^{(s)}) \in \prod_{s\in S} A^{(s)}$,

$$U \subseteq \left\{ s \mid \mathfrak{A}^{(s)} \models \varphi_1\left[h^{(s)}\binom{x}{b^{(s)}}\right]\right\}.$$

The ($\Leftarrow$) direction now follows from filter property (3) (2.6.0.1). To prove ($\Rightarrow$), we assume $U \notin \mathscr{D}$. Then $S \setminus U \in \mathscr{D}$, i.e.

$$V = \left\{ s \mid \mathfrak{A}^{(s)} \not\models \varphi_1\left[h^{(s)}\binom{x}{a^{(s)}}\right] \text{ for some } a^{(s)} \in A^{(s)} \right\} \in \mathscr{D}.$$

Now we define a choice sequence $(b^{(s)})$ by

$$b^{(s)} = \begin{cases} \text{an } a^{(s)} \text{ with } \mathfrak{A}^{(s)} \not\models \varphi_1\left[h^{(s)}\binom{x}{a^{(s)}}\right] & \text{if } s \in V\\ \text{an } a^{(s)} \in A^{(s)} & \text{otherwise.}\end{cases}$$

Then

$$V \subseteq \left\{ s \mid \mathfrak{A}^{(s)} \not\models \varphi_1\left[h^{(s)}\binom{x}{b^{(s)}}\right]\right\}.$$

Since $V \in \mathscr{D}$, the right-hand side also lies in $\mathscr{D}$, i.e.

$$\left\{ s \mid \mathfrak{A}^{(s)} \models \varphi_1\left[h^{(s)}\binom{x}{b^{(s)}}\right]\right\} \notin \mathscr{D}.$$

But this contradicts the hypothesis. Therefore, the proof of the equivalence is complete. □

Now we consider the special case where all $\mathfrak{A}^{(s)}$ are equal to each other; say, $\mathfrak{A}^{(s)} = \mathfrak{B}$ for all $s \in S$. In this case we call

$$\mathfrak{A} = \prod_{s\in S} \mathfrak{B} \Big/ \mathscr{D} =: \mathfrak{B}^S / \mathscr{D}$$

the *ultrapower of* $\mathfrak{B}$ *with respect to* $\mathscr{D}$.

With ultrapowers we have another method to construct elementary extensions:

**Corollary 2.6.3.** *Let* $\mathfrak{B}$ *be an L-structure and D an ultrafilter on S. Then the mapping* $\tau(b) = \overline{(b)_{s\in S}}$ *for* $b \in |\mathfrak{B}|$ *defines an elementary embedding of* $\mathfrak{B}$ *in the ultrapower* $\mathfrak{A} = \mathfrak{B}^S/\mathscr{D}$.

*Proof*: Let $h$ be an evaluation in $\mathfrak{B}$. From $h$ we construct the sequence of evaluations $\overline{h}$ in the ultrapower $\mathfrak{A}$ by setting

$$\overline{h}(x) = \overline{(h(x))_{s\in S}} = \tau(h(x)).$$

Then for every $L$-formula $\varphi$ and every evaluation $h$ in $\mathfrak{B}$ we have, by Theorem 2.6.2,

$$\mathfrak{A} \models \varphi\,[\overline{h}] \quad \text{iff} \quad \{s \mid \mathfrak{B} \models \varphi\,[h]\} \in \mathscr{D}.$$

Since the condition $\mathfrak{B} \models \varphi\,[h]$ does not depend on $s$, the set $\{s \mid \mathfrak{B} \models \varphi\,[h]\}$ can only be empty, or equal to $S$. Since $\emptyset \notin \mathscr{D}$ and $S \in \mathscr{D}$, we therefore obtain

$$\{s \mid \mathfrak{B} \models \varphi\,[h]\} \in \mathscr{D} \quad \text{iff} \quad \mathfrak{B} \models \varphi\,[h].$$

All together we therefore have

$$\mathfrak{A} \models \varphi\,[\tau \circ h] \quad \text{iff} \quad \mathfrak{B} \models \varphi\,[h].$$

But this means exactly that $\tau$ is an elementary embedding of $\mathfrak{B}$ in $\mathfrak{A}$. □

As a first application of ultraproducts, we would like to prove the Finiteness Theorem 1.5.6 (which is fundamental for model theory) without relying on the Completeness Theorem of first-order logic; such a proof was promised in Remark 1.5.7.

For this, let $\Sigma$ be a set of $L$-sentences, every finite subset $\Delta$ of which possesses a model. First we consider the system $S$ of all finite subsets $\Delta$ of $\Sigma$. The set $S$ is partially ordered by set-theoretic inclusion. Next, we define the system $F_0$ of (nonempty) sets

$$S_\Delta = \{\Delta' \in S \mid \Delta \subseteq \Delta'\},$$

where $\Delta$ runs through all elements of $S$. $F_0$ is closed under (finite) intersection, since

$$S_{\Delta_1} \cap S_{\Delta_2} = S_{\Delta_1 \cup \Delta_2}.$$

By Lemma 2.6.1, there is an ultrafilter $\mathscr{D}$ on $S$ that contains $F_0$. We then have:

**Theorem 2.6.4.** *Let* $\Sigma \subseteq \mathrm{Sent}\,L$, *S the system of all finite subsets* $\Delta$ *of* $\Sigma$, *and* $\mathscr{D}$ *an ultrafilter on S that contains* $F_0$. *For every* $\Delta \in S$ *let* $\mathfrak{A}^{(\Delta)}$ *be a model of* $\Delta$. *Then the ultraproduct*

$$\mathfrak{A} = \prod_{\Delta\in S} \mathfrak{A}^{(\Delta)} \Big/ \mathscr{D}$$

*is a model of* $\Sigma$.

*Proof*: Let $\rho \in \Sigma$. For all $\Delta \in S$ with $\rho \in \Delta$, we obviously have $\mathfrak{A}^{(\Delta)} \models \rho$. Therefore

$$\{\Delta \mid \mathfrak{A}^{(\Delta)} \models \rho\} \supseteq \{\Delta \mid \Delta \supseteq \{\rho\}\} =: S_{\{\rho\}}.$$

From $S_{\{\rho\}} \in F_0 \subseteq \mathscr{D}$, it follows that

$$\{\Delta \mid \mathfrak{A}^{(\Delta)} \models \rho\} \in \mathscr{D}.$$

But this implies $\mathfrak{A} \models \rho$, using Łos' Theorem 2.6.2. □

Finally, we would like to discuss the case of the ultrapower

$$\mathfrak{A} = \mathfrak{B}^{\mathbb{N}}/\mathscr{D}$$

of an infinite $L$-structure $\mathfrak{B}$ with respect to an ultrafilter $\mathscr{D}$ on the natural numbers (or on any other countable set). The case of a finite structure $\mathfrak{B}$ is actually uninteresting, since by the elementary equivalence of $\mathfrak{A}$ and $\mathfrak{B}$, both have the same number of elements. Then the embedding $\tau$ given in Corollary 2.6.3 would be an isomorphism of $\mathfrak{B}$ onto $\mathfrak{A}$.

If $\mathscr{D}$ is a principal ultrafilter on $\mathbb{N}$, say,

$$\mathscr{D} = \{U \subseteq \mathbb{N} \mid m \in U\}$$

for a certain $m \in \mathbb{N}$, then $\mathfrak{A}$ is again isomorphic to $\mathfrak{B}$. Indeed, two sequences $\left(a^{(n)}\right)_{n\in\mathbb{N}}$ and $\left(b^{(n)}\right)_{n\in\mathbb{N}}$ of elements of $|\mathfrak{B}|$ are equivalent if and only if $a^{(m)} = b^{(m)}$. Therefore the equivalence classes of sequences $\left(a^{(n)}\right)_{n\in\mathbb{N}}$ correspond bijectively with the elements of $|\mathfrak{B}|$.

If $\mathscr{D}$ is a nonprincipal ultrafilter, then $\mathscr{D}$ cannot contain any finite subset of $\mathbb{N}$, and so it must contain all co-finite subsets of $\mathbb{N}$. Since $B = |\mathfrak{B}|$ is infinite, there is an injective sequence $\left(a^{(n)}\right)_{n\in\mathbb{N}}$ of elements of $B$. The equivalence class $a = \overline{\left(a^{(n)}\right)_{n\in\mathbb{N}}}$ is an element of $\mathfrak{A}$ that does not belong to the image of the embedding $\tau$ described in Corollary 2.6.3. Indeed, if $b \in B$, then from $\tau(b) = a$ it would immediately follow, by the definition of equivalence of sequences, that $D$ would contain either the empty set or a singleton set, since from $\tau(b) = a$ we would have

$$\{n \mid b = a^{(n)}\} \in \mathscr{D}.$$

Therefore in this case the ultrapower $\mathfrak{A}$ is a proper extension of the image $\tau(\mathfrak{B})$.

The following theorem relates ultraproducts with saturation.

**Theorem 2.6.5.** *For each $s \in \mathbb{N}$, let $\mathfrak{A}^{(s)}$ be an $L$-structure, where $L$ is a countable language (i.e. $\kappa_L = \aleph_0$). If $\mathscr{D}$ is a nonprincipal ultrafilter on $\mathbb{N}$, then the ultraproduct*

$$\mathfrak{A} = \prod_{s\in\mathbb{N}} \mathfrak{A}^{(s)} \Big/ \mathscr{D}$$

*is $\aleph_1$-saturated.*

*Proof*: Let $\Phi(v_0)$ be a type of $(\mathfrak{A},A')$ with $\mathrm{card}(A') < \aleph_1$. Thus, the set $\Phi$ is at most countable. Let $(\varphi_n)_{n\in\mathbb{N}}$ be an enumeration of the formulae in $\Phi$ (possibly with repetitions). We construct the sequence $(\psi_n)_{n\in\mathbb{N}}$ by setting

$$\psi_n = (\varphi_0 \wedge \cdots \wedge \varphi_n).$$

If an $a \in A = |\mathfrak{A}|$ realizes all formulae $\psi_n$, then obviously it also realizes all $\varphi_n$.

By hypothesis there is, to each $n \in \mathbb{N}$, an element $a_n \in A$ with $(\mathfrak{A},A) \models \psi_n(\underline{a_n})$. In the formulae $\psi_n$, constant symbols $\underline{b}$ occur, for certain $b \in A'$. If we represent these elements $b$ by sequences $\left(b^{(s)}\right)_{s\in\mathbb{N}}$ and the elements $a_n$ likewise by sequences $\left(a_n^{(s)}\right)_{s\in\mathbb{N}}$, we obtain from Łos' Theorem 2.6.2 (with Lemma 2.4.1) precisely

$$U_n = \left\{ s \in \mathbb{N} \;\middle|\; (\mathfrak{A}^{(s)},A^{(s)}) \models \psi_n^{(s)}\left(\underline{a_n^{(s)}}\right) \right\} \in \mathscr{D},$$

where $\psi_n^{(s)}$ is the result of the replacement of the constant symbols $\underline{b}$ in $\psi_n$ by $\underline{b^{(s)}}$. Since $\mathscr{D}$ is nonprincipal, we also have

$$V_n = \{ s \mid s \geq n \} \cap \bigcap_{i=1}^{n} U_i \in \mathscr{D}. \tag{2.6.5.1}$$

For the set $V_n$ we have

$$V_{n+1} \subseteq V_n \quad \text{and} \quad \bigcap_{n=0}^{\infty} V_n = \emptyset.$$

Now we define $a = \overline{\left(a^{(s)}\right)_{s\in\mathbb{N}}}$ by setting

$$a^{(s)} = \begin{cases} a_n^{(s)} & \text{if } s \in V_n \setminus V_{n+1} \\ c^{(s)} & \text{otherwise.} \end{cases} \tag{2.6.5.2}$$

Here $c^{(s)}$ is a fixed, chosen element of $A^{(s)}$. We claim that the equivalence class $a$ satisfies all formulae $\psi_n$ in $\mathfrak{A}$. For this it suffices, by Łos' Theorem 2.6.2, to show

$$W_n = \left\{ s \;\middle|\; (\mathfrak{A}^{(s)},A^{(s)}) \models \psi_n^{(s)}\left(a^{(s)}\right) \right\} \in \mathscr{D}.$$

However, since $\psi_m^{(s)}$ implies $\psi_n^{(s)}$ for $m \geq n$, it follows immediately from (2.6.5.2) that

$$V_m \setminus V_{m+1} \subseteq W_n$$

for each $m \geq n$. Therefore $W_n$ contains the union of all $V_m \setminus V_{m+1}$ for $m \geq n$, which, in view of $\bigcap_{m\in\mathbb{N}} V_m = \emptyset$, equals $V_n$. Then $W_n \in \mathscr{D}$ follows from (2.6.5.1). □

## 2.7 Exercises for Chapter 2

**2.7.1.** An *ordering* of a group $G$ is a linear ordering $<$ on (the underlying set of) $G$ that satisfies, in addition,

$$a < b \rightarrow ac < bc \wedge ca < cb,$$

for all $a, b, c \in G$.

(a) Show that a group $G$ can be ordered if each finitely generated subgroup of $G$ can be ordered.

*Hint*: Use the Diagram Lemma 2.4.2 and the Finiteness Theorem 1.5.6.

(b) Every Abelian group $G$ is orderable if and only if it is torsion free.

*Hint*: Use the Fundamental Theorem (or "Structure Theorem") on finitely generated Abelian groups.

**2.7.2.** Let $L = (\cdot, 1)$ be the language of (multiplicatively written) groups. We call a group $G$ a *torsion group* if for every $a \in G$ there is an $n \in \mathbb{N}$ with $a^n = 1$.

Show that the class of torsion groups cannot be axiomatised[4] in $L$.

**2.7.3.** (a) Let $L_{\geq} = (\geq)$ be the language of ordered sets. Show that the class of well-ordered sets cannot be axiomatized[4] in $L_{\geq}$.

(b) Let $L_R = (+, \cdot, 0, 1)$ be the language of rings. Show that the class of fields of characteristic 0 cannot be finitely axiomatized[4] in $L_R$.

*Hint*: Consider a suitable ultraproduct of all the fields $\mathbb{F}_p$ ($p$ a prime number).

**2.7.4. Definitions:** Let $L$ be a (first-order) language, and let $\mathfrak{A}$ and $\mathfrak{B}$ be two $L$-structures with universes $A$ and $B$, respectively.

(a) A function $p\colon \mathrm{dom}(p) \rightarrow \mathrm{im}(p)$ is a *partial isomorphism* of $\mathfrak{A}$ into $\mathfrak{B}$ if:

1. $\mathrm{dom}(p) \subseteq A$ and $\mathrm{im}(p) \subseteq B$,
2. $p$ is bijective,
3. for each $n$-place relation symbol $R$ in $L$, and all $a_1, \ldots, a_n \in \mathrm{dom}(p)$,
   $R^{\mathfrak{A}}(a_1, \ldots, a_n)$ if and only if $R^{\mathfrak{B}}(p(a_1), \ldots, p(a_n))$,
4. for each $n$-place function symbol $f$ in $L$, and all $a_1, \ldots, a_n, a \in \mathrm{dom}(p)$,
   $f^{\mathfrak{A}}(a_1, \ldots, a_n) = a$ if and only if $f^{\mathfrak{B}}(p(a_1), \ldots, p(a_n)) = p(a)$, and
5. for each constant symbol $c$ in $L$, and all $a \in \mathrm{dom}(p)$,
   $c^{\mathfrak{A}} = a$ if and only if $c^{\mathfrak{B}} = p(a)$.

(b) Two structures $\mathfrak{A}$ and $\mathfrak{B}$ are called *partially isomorphic* ($\mathfrak{A} \cong_p \mathfrak{B}$) if there is a nonempty set $I$ of partial isomorphisms from $\mathfrak{A}$ into $\mathfrak{B}$ with the following properties:

1. For each $p \in I$ and each $a \in A$, there is an extension $q$ of $p$ in $I$ with $a \in \mathrm{dom}(q)$.
2. For each $p \in I$ and each $b \in B$, there is an extension $q$ of $p$ in $I$ with $b \in \mathrm{im}(q)$.

**Exercises:**

(c) Give a partial isomorphism with nonempty domain from the additive group $(\mathbb{R}, +, 0)$ into the additive group $(\mathbb{Z}, +, 0)$.

---

[4] Recall the definition of (finitely) "axiomatized" given in Exercise 1.7.14.

(d) Let $L$ be the empty language. Show that every two infinite $L$-structures are partially isomorphic.

(e) Give a language $L$, and partially isomorphic $L$-structures $\mathfrak{A}$ and $\mathfrak{B}$ that are not isomorphic.

**2.7.5.** Let $L$ be a language containing only relation symbols, and let $\mathfrak{A}$ and $\mathfrak{B}$ be two partially isomorphic $L$-structures. Show that $\mathfrak{A}$ and $\mathfrak{B}$ are then also elementarily equivalent, i.e. each $L$-sentence holds in $\mathfrak{A}$ if and only if it holds in $\mathfrak{B}$.

*Hint*: If $I$ is as in 2.7.4(b) above, then for each $L$-formula $\varphi(x_1,\dots,x_n)$, each $p \in I$, and all $a_1,\dots,a_n \in \mathrm{dom}(p)$, show that $\mathfrak{A} \models \varphi(a_1,\dots,a_n)$ if and only if $\mathfrak{B} \models \varphi(p(a_1),\dots,p(a_n))$.

**2.7.6.** Show that the linearly ordered sets $\langle \mathbb{R}, \le \rangle$ and $\langle \mathbb{Q}, \le \rangle$ are elementarily equivalent.

**2.7.7.** Let $S$ be a nonempty set, and let $\mathscr{F}$ be a filter on $S$. Show that the following sentences are equivalent:

(a) $\mathscr{F}$ is an ultrafilter on $S$.

(b) For all $U,V \subset S$, if $U \cup V \in \mathscr{F}$, then either $U \in \mathscr{F}$ or $V \in \mathscr{F}$.

(c) $\mathscr{F}$ is a maximal filter on $S$.

Show, in addition, that every filter on $S$ is contained in an ultrafilter on $S$.

**2.7.8.** Let $S$ be an infinite set, and let $\mathscr{F}_0 := \{U \subset S \mid S \setminus U \text{ is finite}\}$. Show:

(a) $\mathscr{F}_0$ is a filter on $S$, but not an ultrafilter.

(b) An ultrafilter on $S$ is a principal ultrafilter if and only if it does not contain $\mathscr{F}_0$.

(c) There is an ultrafilter on $S$ that is not a principal ultrafilter.

**2.7.9.** Let $K^{(s)}$ (for $s \in S$) be a family of fields, and $R = \prod_{s\in S} K^{(s)}$ the direct product of the $K^{(s)}$. If $\mathscr{D}$ is an ultrafilter on $S$, let

$$M_{\mathscr{D}} = \left\{ \left(a^{(s)}\right)_{s\in S} \in R \,\middle|\, \left\{ s \,\middle|\, a^{(s)} = 0 \right\} \in \mathscr{D} \right\}.$$

Show:

(a) $M_{\mathscr{D}}$ is a maximal ideal of $R$.

(b) For every maximal ideal $M$ of $R$, there is an ultrafilter $\mathscr{D}$ on $S$ with $M = M_{\mathscr{D}}$.

(c) The ultraproduct $R = \prod_{s\in S} K^{(s)} \big/ \mathscr{D}$ is isomorphic to the residue field $R/M_{\mathscr{D}}$.

**2.7.10.** Let $\mathfrak{R}^*$ be the ordered field $\langle \mathbb{R}, +, \cdot, 0, 1, \le \rangle^{\mathbb{N}} / \mathscr{F}$, where $\mathscr{F}$ is a nonprincipal ultrafilter on $\mathbb{N}$. Show:

(a) The mapping $r \mapsto \overline{(r)_{n\in\mathbb{N}}}$ (using the notation (2.6.1.3)) defines an elementary embedding of $\mathfrak{R} := \langle \mathbb{R}, +, \cdot, 0, 1, \le \rangle$ into $\mathbb{R}^*$.

(b) The set $\mathscr{O} := \{a \in \mathbb{R}^* \mid \exists r \in \mathbb{R}\,(r > 0 \wedge -r \le a \le r)\}$ is a convex subring of $\mathbb{R}^*$. Here a subset $M$ of $\mathbb{R}^*$ is called *convex* if for all $a,b \in \mathbb{R}^*$, $0 \le a \le b$ and $b \in M$ imply $a \in M$. We call the elements of $\mathscr{O}$ *finite*.

(c) The set $\mathscr{M} := \{a \in \mathbb{R}^* \mid \forall r \in \mathbb{R}\,(r > 0 \Rightarrow -r \le a \le r)\}$ is a convex ideal in $\mathscr{O}$. We call the elements of $\mathfrak{M}$ *infinitely small* or *infinitesimal*.

(d) For each finite element $a \in \mathbb{R}^*$ there is exactly one real number $r \in \mathbb{R}$ with $r - a \in \mathfrak{M}$. We call this real number the *standard part* of $a$.

(e) A function $f\colon \mathbb{R} \to \mathbb{R}$ is continuous at a point $x \in \mathbb{R}$ if and only if $f^*(x+h) - f(x) \in \mathfrak{M}$ for every infinitesimal $h \in \mathbb{R}^*$. Here $f^*$ is the extension of $f$ from $\mathbb{R}$ to $\mathbb{R}^*$ defined by $f^*\left(\overline{(a_n)_{n\in\mathbb{N}}}\right) := \overline{(f(a_n))_{n\in\mathbb{N}}}$.

**2.7.11.** Show that the Abelian group $\mathbb{Z} = \langle \mathbb{Z};\, +^{\mathbb{Z}};\, 0^{\mathbb{Z}} \rangle$ is not a direct summand of the extension group $\mathbb{Z}^* = \mathbb{Z}^{\mathbb{N}} / \mathscr{D}$, where $\mathscr{D}$ is a nonprincipal ultrafilter on $\mathbb{N}$.

*Hint*: Consider the formula $\varphi_2(v_0)$ expressing the congruence $v_0 \equiv 1 \pmod 2$, and, for each prime $p > 2$, the formula $\varphi_p(v_0)$ expressing the congruence $v_0 \equiv 0 \pmod p$. Pay attention to Theorem 2.6.5.

**2.7.12.** Let $L$ be a formal language, and $S$ an arbitrary set. For each $s \in S$, let $\mathfrak{A}^{(s)}$ be an $L$-structure. Let $s_0 \in S$ be given, and let $\mathscr{H}(s_0)$ be the principal ultrafilter on $s_0$ (recall the notation (2.6.0.2)). Show that

$$\prod_{s\in S} \mathfrak{A}^{(s)} \Big/ \mathscr{H}(s_0) \cong \mathfrak{A}^{(s_0)}.$$

**2.7.13.** Show that the existence of a saturated, divisible, ordered Abelian group of cardinality $\aleph_1$ is equivalent to the ("special") Continuum Hypothesis $2^{\aleph_0} = \aleph_1$.

**2.7.14.** Most arguments with saturated structures (whose existence usually presupposes the Generalized Continuum Hypothesis (GCH), as explained in Remark 2.5.3) can instead be carried out just as well with "special structures", whose existence does not require GCH. Here we call an infinite structure $\mathfrak{A}$ *special* if it is the union of an elementary chain $(\mathfrak{A}_\alpha)_{\alpha<\kappa}$ with $\kappa = \mathrm{card}(|\mathfrak{A}|)$, where $\alpha$ ranges over *cardinal* numbers and $\mathfrak{A}_\alpha$ is $\alpha^+$-saturated. Show:

(a) *Existence*: To each cardinal number $\kappa > \kappa_L$ with $\kappa = \sum_{\alpha<\kappa} 2^\alpha$, and to each infinite structure $\mathfrak{A}$ with $\mathrm{card}(|\mathfrak{A}|) < \kappa$, there is a special elementary extension $\mathfrak{A}^*$ of $\mathfrak{A}$ with $\mathrm{card}(|\mathfrak{A}^*|) = \kappa$.

(b) There are arbitrarily large infinite cardinal numbers $\kappa$ with $\kappa = \sum_{\alpha<\kappa} 2^\alpha$.

(c) *Uniqueness*: Two infinite special $L$-structures $\mathfrak{A}$ and $\mathfrak{A}'$ of equal cardinality $\kappa$ are isomorphic if and only if they are elementarily equivalent.

**2.7.15.** Let $L = (+, \cdot, 0, 1)$. For $L$-structures $\mathfrak{A}$ and $\mathfrak{A}'$, we write $\mathfrak{A} \not\equiv \mathfrak{A}'$ if $\mathfrak{A}$ and $\mathfrak{A}'$ are not elementarily equivalent. Show:

(i) $\langle \mathbb{Q}, +, \cdot, 0, 1 \rangle \not\equiv \langle \mathbb{Q}(\sqrt{2}), +, \cdot, 0, 1 \rangle$

(ii) $\langle \mathbb{R}, +, \cdot, 0, 1 \rangle \not\equiv \langle \mathbb{R}(X), +, \cdot, 0, 1 \rangle$

(iii) $\langle \mathbb{R}[X], +, \cdot, 0, 1 \rangle \not\equiv \langle \mathbb{R}[X,Y], +, \cdot, 0, 1 \rangle$

# Chapter 3
# Properties of Model Classes

In this chapter we wish to study properties of model classes. By a *model class* we mean the class of all models of an axiom system $\Sigma$.

First we shall furnish such a class with a topology whose compactness is exactly the content of the Finiteness Theorem 1.5.6. After that, we shall introduce several properties of the model class of an axiom system $\Sigma$, whose study can lead, among other things, to the proof of the completeness (recall §1.6) of $\Sigma$. We shall carry this out explicitly for a series of theories axiomatized in (1.6).

The properties of $\Sigma$ or of its model class to be introduced are: categoricity in a fixed cardinality, model completeness and quantifier elimination. The study of these properties is based not only on the possible applicability to the proof of the completeness of a theory, but is, rather, also justified by its usefulness in concrete, mathematical (in particular, algebraic) theories. In this chapter we shall investigate such properties only for the theory of algebraically closed fields; in Chapter 4, other theories will follow.

## 3.1 Compactness and Separation

First we want to interpret the statement of the Finiteness Theorem 1.5.6 as the compactness of a certain space. This translation is not really profound, but very suggestive.

We consider a language $L = (\lambda, \mu, K)$ and denote by $\mathrm{Mod}_L$ the class of all $L$-structures $\mathfrak{A}$. For a subset $\Sigma$ of $\mathrm{Sent}(L)$, we define

$$\mathrm{Mod}_L(\Sigma) = \{\mathfrak{A} \in \mathrm{Mod}_L \mid \mathfrak{A} \models \Sigma\}.$$

Here, as in Chapter 2, the notation $\mathfrak{A} \models \Sigma$ means nothing other than that $\mathfrak{A} \models \sigma$ for all $\sigma \in \Sigma$. The class $\mathrm{Mod}_L(\Sigma)$ is the class of all models of $\Sigma$. It will therefore also be called the *model class* of $\Sigma$. For $\Sigma = \{\sigma\}$, we write, briefly, $\mathrm{Mod}_L(\sigma)$, instead of $\mathrm{Mod}_L(\{\sigma\})$. We shall also drop the subscript $L$ whenever it is clear which language

A. Prestel, C.N. Delzell, *Mathematical Logic and Model Theory*, Universitext,
DOI 10.1007/978-1-4471-2176-3_4, © Springer-Verlag London Limited 2011

$L$ we are dealing with. The following properties are immediately verified:

$$\begin{aligned}
&\mathrm{Mod}(\exists x\, x \neq x) = \emptyset \\
&\mathrm{Mod}(\forall x\, x \doteq x) = \mathrm{Mod} \\
&\mathrm{Mod}(\neg\varphi) = \mathrm{Mod} \setminus \mathrm{Mod}(\varphi) \\
&\mathrm{Mod}(\varphi \wedge \psi) = \mathrm{Mod}(\varphi) \cap \mathrm{Mod}(\psi) \\
&\mathrm{Mod}(\varphi \vee \psi) = \mathrm{Mod}(\varphi) \cup \mathrm{Mod}(\psi) \\
&\mathrm{Mod}(\Sigma) = \bigcap_{\sigma \in \Sigma} \mathrm{Mod}(\sigma).
\end{aligned}$$

The system of classes $\mathrm{Mod}(\sigma)$ with $\sigma \in \Sigma$ is therefore closed under intersection, and can therefore serve as a basis of a topology on Mod. We call a subclass of Mod *open* if it is an arbitrary union of classes of the form $\mathrm{Mod}(\sigma)$ with $\sigma \in \mathrm{Sent}(L)$; that is, open classes are of the form

$$\bigcup_{\varphi \in \Phi} \mathrm{Mod}(\varphi),$$

where $\Phi$ is a subset of $\mathrm{Sent}(L)$. The closed classes are then the complements of open classes of the form

$$\bigcap_{\varphi \in \Phi} \mathrm{Mod}(\neg\varphi),$$

i.e. they are just model classes like those we introduced above. (Set $\Sigma = \{\neg\varphi \mid \varphi \in \Phi\}$.) Closed subclasses are often called *elementary* or *axiomatizable*.

*Remark 3.1.1* (topologies on proper classes versus sets). In the sense of set theory, $\mathrm{Mod}_L$ is a proper class. The introduction of a topology on $\mathrm{Mod}_L$ may therefore seem somewhat suspicious. However, we do not wish to dwell on this point. In fact, this defect is easily removed by a set-theoretically precise procedure. One can restrict, in the case of $\mathrm{Mod}_L$, for example, to those $L$-structures that lie in a prescribed set $V$. Here, one chooses the "universe" $V$ large enough so that all interesting set-theoretic operations can be carried out in it.

Consequently, we shall no longer distinguish between sets and classes.

As a first observation, we wish to note that Mod, together with the topology just introduced on it, is not a Hausdorff space. We see this immediately upon reformulating the condition $\mathfrak{A} \equiv \mathfrak{B}$ for $L$-structures $\mathfrak{A}$ and $\mathfrak{B}$:

$$(\mathfrak{A} \in \mathrm{Mod}(\sigma) \text{ iff } \mathfrak{B} \in \mathrm{Mod}(\sigma)) \quad \text{for all } \sigma \in \mathrm{Sent}(L).$$

Therefore, if $\mathfrak{A} \equiv \mathfrak{B}$, then $\mathfrak{B}$ lies in every open set of the form $\mathrm{Mod}(\sigma)$ in which $\mathfrak{A}$ lies, so that $\mathfrak{A}$ can not be separated from $\mathfrak{B}$. But a pair of different $L$-structures $\mathfrak{A}, \mathfrak{B}$ with $\mathfrak{A} \equiv \mathfrak{B}$ that are not even isomorphic is easily constructed: Let $\mathfrak{A}$ be any infinite

$L$-structure, and let $\mathfrak{B}$ be elementarily equivalent to $\mathfrak{A}$ and of higher cardinality than $\mathfrak{A}$ (such a $\mathfrak{B}$ exists by Corollary 2.1.2); then obviously $\mathfrak{A} \not\cong \mathfrak{B}$ and $\mathfrak{A} \equiv \mathfrak{B}$.

**Theorem 3.1.2** (Compactness Theorem). *Let $\Sigma \subseteq \mathrm{Sent}(L)$. Then the "Heine–Borel covering theorem" holds for $\mathrm{Mod}_L(\Sigma)$; i.e. if $\bigcup_{\varphi\in\Phi} \mathrm{Mod}(\varphi)$ is a covering of $\mathrm{Mod}_L(\Sigma)$, then there exist $\varphi_1,\ldots,\varphi_n \in \Phi$ with*

$$\mathrm{Mod}_L(\Sigma) \subseteq \mathrm{Mod}_L(\varphi_1) \cup \cdots \cup \mathrm{Mod}_L(\varphi_n). \tag{3.1.2.1}$$

*Proof*: From the hypothesis that

$$\bigcap_{\sigma\in\Sigma} \mathrm{Mod}(\sigma) \subseteq \bigcup_{\varphi\in\Phi} \mathrm{Mod}(\varphi),$$

it follows immediately that

$$\bigcap_{\sigma\in\Sigma} \mathrm{Mod}(\sigma) \cap \bigcap_{\varphi\in\Phi} \mathrm{Mod}(\neg\varphi) = \emptyset,$$

i.e. the set $\Sigma \cup \{\neg\varphi \mid \varphi \in \Phi\}$ has no model. By the Finiteness Theorem 1.5.6, there must exist a finite subset of $\Sigma \cup \{\neg\varphi \mid \varphi \in \Phi\}$ that possesses no model. In particular, there therefore exist $\varphi_1,\ldots,\varphi_n \in \Phi$ such that

$$\mathrm{Mod}(\Sigma) \cap \mathrm{Mod}(\neg\varphi_1) \cap \cdots \cap \mathrm{Mod}(\neg\varphi_n) = \emptyset.$$

But this means exactly (3.1.2.1). □

The fact that, conversely, the Compactness Theorem implies the Finiteness Theorem (mentioned above), is immediately seen as follows: if $\Sigma \subseteq \mathrm{Sent}(L)$ has no model, then $\bigcap_{\sigma\in\Sigma} \mathrm{Mod}(\sigma) = \emptyset$, whence $\bigcup_{\sigma\in\Sigma} \mathrm{Mod}(\neg\sigma)$ is a covering of Mod. By the Compactness Theorem, there exist, then, $\sigma_1,\ldots,\sigma_n \in \Sigma$ such that already $\mathrm{Mod}(\neg\sigma_1) \cup \cdots \cup \mathrm{Mod}(\neg\sigma_n)$ covers Mod. Therefore

$$\mathrm{Mod}(\sigma_1) \cap \cdots \cap \mathrm{Mod}(\sigma_n) = \emptyset,$$

i.e. the finite subset $\{\sigma_1,\ldots,\sigma_n\}$ of $\Sigma$ possesses no model.

From the Compactness Theorem we now derive the Separation Lemma:

**Lemma 3.1.3** (Separation Lemma). *Let $\Sigma_1, \Sigma_2, \Gamma \subseteq \mathrm{Sent}(L)$ with $\mathrm{Mod}_L(\Sigma_i) \neq \emptyset$ for $i = 1,2$. Suppose that to each $\mathfrak{A} \in \mathrm{Mod}_L(\Sigma_1)$ and to each $\mathfrak{B} \in \mathrm{Mod}_L(\Sigma_2)$ there is a $\gamma \in \Gamma$ with $\mathfrak{A} \models \gamma$ and $\mathfrak{B} \models \neg\gamma$. Then there exists a $\gamma^*$ with*

$$\mathrm{Mod}_L(\Sigma_1) \subseteq \mathrm{Mod}_L(\gamma^*) \quad \textit{and} \quad \mathrm{Mod}_L(\Sigma_2) \subseteq \mathrm{Mod}_L(\neg\gamma^*),$$

*where $\gamma^*$ is a finite disjunction of finite conjunctions of elements of $\Gamma$.*

*Proof*: First we fix an $\mathfrak{A} \in \mathrm{Mod}(\Sigma_1)$, and for each $\mathfrak{B} \in \mathrm{Mod}(\Sigma_2)$ we choose a $\gamma_{\mathfrak{B}}$ with $\mathfrak{A} \in \mathrm{Mod}(\gamma_{\mathfrak{B}})$ and $\mathfrak{B} \in \mathrm{Mod}(\neg\gamma_{\mathfrak{B}})$. Then the classes $\mathrm{Mod}(\neg\gamma_{\mathfrak{B}})$ obviously form a(n open) cover of $\mathrm{Mod}(\Sigma_2)$. By the Compactness Theorem there are

$\mathfrak{B}_1,\ldots,\mathfrak{B}_m \in \mathrm{Mod}(\Sigma_2)$ such that

$$\mathrm{Mod}(\neg\gamma_{\mathfrak{B}_1})\cup\cdots\cup\mathrm{Mod}(\neg\gamma_{\mathfrak{B}_m}) = \mathrm{Mod}(\neg(\gamma_{\mathfrak{B}_1}\wedge\cdots\wedge\gamma_{\mathfrak{B}_m}))$$

already contains $\mathrm{Mod}(\Sigma_2)$. If we set $\gamma_{\mathfrak{A}} = (\gamma_{\mathfrak{B}_1}\wedge\cdots\wedge\gamma_{\mathfrak{B}_m})$, then

$$\mathfrak{A} \in \mathrm{Mod}(\gamma_{\mathfrak{A}}) \quad \text{and} \quad \mathrm{Mod}(\Sigma_2) \subseteq \mathrm{Mod}(\neg\gamma_{\mathfrak{A}}).$$

But now, on the other hand, the classes $\mathrm{Mod}(\gamma_{\mathfrak{A}})$ with this property obviously cover the class $\mathrm{Mod}(\Sigma_1)$. Again, this cover admits a finite subcover; i.e. there exist $\mathfrak{A}_1,\ldots,\mathfrak{A}_n \in \mathrm{Mod}(\Sigma_1)$ such that

$$\mathrm{Mod}(\Sigma_1) \subseteq \mathrm{Mod}(\gamma_{\mathfrak{A}_1})\cup\cdots\cup\mathrm{Mod}(\gamma_{\mathfrak{A}_n}) = \mathrm{Mod}(\gamma_{\mathfrak{A}_1}\vee\cdots\vee\gamma_{\mathfrak{A}_n}).$$

At the same time we have, however,

$$\mathrm{Mod}(\Sigma_2) \subseteq \mathrm{Mod}(\neg\gamma_{\mathfrak{A}_1})\cap\cdots\cap\mathrm{Mod}(\neg\gamma_{\mathfrak{A}_n}) = \mathrm{Mod}(\neg(\gamma_{\mathfrak{B}_1}\vee\cdots\vee\gamma_{\mathfrak{A}_n})).$$

Therefore, if we finally set $\gamma^* = (\gamma_{\mathfrak{A}_1}\vee\cdots\vee\gamma_{\mathfrak{A}_n})$, then we shall have completed the proof. □

As an application, we characterize the finitely axiomatizable subclasses of Mod, i.e. the model classes of finite sets $\{\varphi_1,\ldots,\varphi_n\} \subseteq \mathrm{Sent}(L)$. By the intersection-closedness, these are just the classes of the form $\mathrm{Mod}(\varphi)$ for a $\varphi \in \mathrm{Sent}(L)$.

**Corollary 3.1.4.** *A subclass* $\mathrm{Mod}_L(\Sigma)$ *of* $\mathrm{Mod}_L$ *is finitely axiomatizable if and only if* $\mathrm{Mod}_L \setminus \mathrm{Mod}_L(\Sigma)$ *is axiomatizable.*

*Proof*: We exclude the trivial cases $\mathrm{Mod}(\Sigma) = \mathrm{Mod}$ or $= \emptyset$. If $\mathrm{Mod}(\Sigma) = \mathrm{Mod}(\varphi)$ for some $\varphi \in \mathrm{Sent}(L)$, then naturally $\mathrm{Mod}\setminus\mathrm{Mod}(\Sigma) = \mathrm{Mod}(\neg\varphi)$. If, conversely, $\mathrm{Mod}\setminus\mathrm{Mod}(\Sigma) = \mathrm{Mod}(\Sigma_2)$ for some subset $\Sigma_2 \subseteq \mathrm{Sent}(L)$, then we apply the Separation Lemma to $\Gamma = \Sigma_1 = \Sigma$. If $\mathfrak{B} \in \mathrm{Mod}(\Sigma_2)$, then $\mathfrak{B}$ is not a model of $\Sigma$, whence there is a $\gamma \in \Sigma$ with $\mathfrak{B} \models \neg\gamma$. On the other hand, $\mathfrak{A} \models \gamma$ naturally holds for all $\mathfrak{A} \in \mathrm{Mod}(\Sigma_1)$. Therefore the hypothesis of the Separation Lemma is satisfied, as a consequence of which there is a sentence $\gamma^*$ of the form

$$(\gamma_{11}\wedge\cdots\wedge\gamma_{1m_1})\vee\cdots\vee(\gamma_{n1}\wedge\cdots\wedge\gamma_{nm_n}),$$

with $\gamma_{ij} \in \Sigma$, such that $\mathrm{Mod}(\Sigma) \subseteq \mathrm{Mod}(\gamma^*)$ and $\mathrm{Mod}(\Sigma_2) \subseteq \mathrm{Mod}(\neg\gamma^*)$. But this last means exactly that $\mathrm{Mod}(\gamma^*) \subseteq \mathrm{Mod}(\Sigma)$. Therefore $\mathrm{Mod}(\Sigma) = \mathrm{Mod}(\gamma^*)$. □

*Remark 3.1.5.* We remark, in addition, that a class $\mathrm{Mod}(\Sigma)$ that is finitely axiomatizable can be axiomatized already by finitely many axioms from $\Sigma$. Namely, with the notation of the last proof, we even have

$$\mathrm{Mod}(\Sigma) = \mathrm{Mod}(\{\gamma_{ij} \mid 1 \le j \le m_i,\ 1 \le i \le n\}).$$

One may easily check this.

The following lemma is a consequence of the Separation Lemma that is useful for applications. For this we wish to use a generalization of a notation introduced in Theorem 2.5.4. For $L$-structures $\mathfrak{A}, \mathfrak{B}$ and for $\Gamma \subseteq \mathrm{Sent}(L)$,

$$\mathfrak{A} \overset{\Gamma}{\rightsquigarrow} \mathfrak{B}$$

will mean that for all $\gamma \in \Gamma$,

$$\mathfrak{A} \models \gamma \quad \text{implies} \quad \mathfrak{B} \models \gamma.$$

For $\mathfrak{A} \overset{\{\varphi\}}{\rightsquigarrow} \mathfrak{B}$ we write, briefly, $\mathfrak{A} \overset{\varphi}{\rightsquigarrow} \mathfrak{B}$.

**Lemma 3.1.6.** *Let $\Sigma \cup \Gamma \cup \{\varphi\} \subseteq \mathrm{Sent}(L)$ and let $\gamma_0$ (respectively, $\gamma_1$) be an element of $\Gamma$ that holds in no model (respectively, in all models) of $\Sigma$. Suppose, in addition, that for all models $\mathfrak{A}, \mathfrak{B}$ of $\Sigma$, whenever $\mathfrak{A} \overset{\Gamma}{\rightsquigarrow} \mathfrak{B}$, then also $\mathfrak{A} \overset{\varphi}{\rightsquigarrow} \mathfrak{B}$. Then there exists a finite disjunction $\gamma^*$ of finite conjunctions of elements of $\Gamma$ that is equivalent to $\varphi$ in all models of $\Sigma$; i.e. $\Sigma \vdash (\varphi \leftrightarrow \gamma^*)$.*[1]

*Proof*: In the Separation Lemma we set $\Sigma_1 = \Sigma \cup \{\varphi\}$ and $\Sigma_2 = \Sigma \cup \{\neg\varphi\}$. If $\mathrm{Mod}(\Sigma_1) = \emptyset$, then obviously $\Sigma \vdash (\varphi \leftrightarrow \gamma_0)$; if $\mathrm{Mod}(\Sigma_2) = \emptyset$, then, analogously, $\Sigma \vdash (\varphi \leftrightarrow \gamma_1)$. Now if $\mathfrak{A} \in \mathrm{Mod}(\Sigma_1)$ and $\mathfrak{B} \in \mathrm{Mod}(\Sigma_2)$, then $\mathfrak{A} \overset{\varphi}{\rightsquigarrow} \mathfrak{B}$ cannot hold, and hence neither can $\mathfrak{A} \overset{\Gamma}{\rightsquigarrow} \mathfrak{B}$. Accordingly, there exists a $\gamma \in \Gamma$ with $\mathfrak{A} \models \gamma$ and $\mathfrak{B} \models \neg\gamma$. Then, by the Separation Lemma, we obtain a finite disjunction $\gamma^*$ of finite conjunctions of elements of $\Gamma$ with $\mathrm{Mod}(\Sigma_1) \subseteq \mathrm{Mod}(\gamma^*)$ and $\mathrm{Mod}(\Sigma_2) \subseteq \mathrm{Mod}(\neg\gamma^*)$. From this it follows, for $\mathfrak{A} \in \mathrm{Mod}(\Sigma)$, that, on the one hand,

$$\mathfrak{A} \models \varphi \quad \text{implies} \quad \mathfrak{A} \models \gamma^*,$$

and, on the other hand,

$$\mathfrak{A} \models \neg\varphi \quad \text{implies} \quad \mathfrak{A} \models \neg\gamma^*.$$

Therefore, for all $\mathfrak{A} \in \mathrm{Mod}(\Sigma)$ we also have

$$\mathfrak{A} \models (\varphi \leftrightarrow \gamma^*). \qquad \square$$

As an application of Lemma 3.1.6, we wish to prove the following theorem:

**Theorem 3.1.7.** *Let $\Sigma \cup \{\varphi\} \subseteq \mathrm{Sent}(L)$. Suppose that for all $\mathfrak{B}, \mathfrak{B}_1 \in \mathrm{Mod}(\Sigma)$ with $\mathfrak{B} \subseteq \mathfrak{B}_1$, $\mathfrak{B}_1 \overset{\varphi}{\rightsquigarrow} \mathfrak{B}$ holds. Then there is a quantifier free $L$-formula $\delta$ with $\Sigma \vdash (\varphi \leftrightarrow \forall\delta)$.*

*Proof*: We set $\Gamma = \{\forall\delta \mid \delta \in \mathrm{Fml}(L),\ \delta \text{ quantifier-free}\}$ and show, for $\mathfrak{A}, \mathfrak{B} \in \mathrm{Mod}(\Sigma)$, that $\mathfrak{A} \overset{\Gamma}{\rightsquigarrow} \mathfrak{B}$ always implies $\mathfrak{A} \overset{\varphi}{\rightsquigarrow} \mathfrak{B}$. Toward this purpose, we observe

[1] Recall that by Gödel's Completeness Theorem 1.5.2, $\psi$ holds in every model of $\Sigma$ if and only if $\Sigma \vdash \psi$.

that the hypothesis $\mathfrak{A} \overset{\Gamma}{\rightsquigarrow} \mathfrak{B}$, when read backwards, says that every $\exists$-sentence that holds in $\mathfrak{B}$ holds already in $\mathfrak{A}$. If we now choose (by Theorem 2.5.2) a $\kappa$-saturated elementary extension $\mathfrak{A}'$ of $\mathfrak{A}$ with $\kappa \geq \text{card}(|\mathfrak{B}|)$, then Theorem 2.5.4 yields an embedding of $\mathfrak{B}$ into $\mathfrak{A}'$. Then by our hypothesis, $\mathfrak{B} \models \varphi$ will follow from $\mathfrak{A}' \models \varphi$.

Now we apply Lemma 3.1.6 with the $\gamma_0$ there being the sentence $\forall v_0\ v_0 \neq v_0$, and with $\gamma_1$ being the sentence $\forall v_0\ v_0 \doteq v_0$.

Then the conclusion of Lemma 3.1.6 is that $\Sigma \vdash (\varphi \leftrightarrow \gamma^*)$, where $\gamma^*$ is a disjunction of conjunctions from $\Gamma$. Now one sees easily that for the above choice of $\Gamma$, $\gamma^*$ is purely logically equivalent to some $\gamma \in \Gamma$, i.e. $\vdash (\gamma^* \leftrightarrow \gamma)$ holds. Hence the theorem is proved. □

## 3.2 Categoricity

Now we would like to begin to prove the completeness of several theories that we introduced in §1.6. We called (1.6.0.2) an axiom system $\Sigma \subseteq \text{Sent}(L)$ *complete* if for all $\sigma \in \text{Sent}(L)$, either $\Sigma \vdash \sigma$ or $\Sigma \vdash \neg\sigma$. For an $L$-theory $T$, this means that

$$\sigma \in T \quad \text{or} \quad \neg\sigma \in T.$$

Already in (1.6.0.3), we had seen that complete $L$-theories $T$ are always of the form $\text{Th}(\mathfrak{A})$, for an $L$-structure $\mathfrak{A}$. A further, trivial criterion is:

**Lemma 3.2.1.** *$\Sigma \subseteq \text{Sent}(L)$ is complete if and only if every two models of $\Sigma$ are elementarily equivalent.*

*Proof*: If $\Sigma$ is complete, then $\text{Ded}(\Sigma) = \text{Th}(\mathfrak{A})$ for an $L$-structure $\mathfrak{A}$, and if $\mathfrak{B}$ is a model of $\Sigma$, then, naturally, $\mathfrak{B} \models \text{Th}(\mathfrak{A})$. From this, $\mathfrak{A} \equiv \mathfrak{B}$ follows immediately. If $\mathfrak{C}$ is a further model of $\Sigma$, then $\mathfrak{A} \equiv \mathfrak{C}$ follows analogously. By transitivity, $\mathfrak{B} \equiv \mathfrak{C}$.

If $\Sigma$ is incomplete, then there exists a $\sigma \in \text{Sent}(L)$ with $\Sigma \nvdash \sigma$ and $\Sigma \nvdash \neg\sigma$. By Gödel's Completeness Theorem 1.5.2, then, there are models $\mathfrak{A}$ and $\mathfrak{B}$ of $\Sigma$ with $\mathfrak{A} \models \neg\sigma$ and $\mathfrak{B} \models \sigma$. But this contradicts the hypothesis that $\mathfrak{A} \equiv \mathfrak{B}$. □

A very simple, sufficient condition for completeness, which, however, can be applied only very seldomly, is:

**Theorem 3.2.2** (Vaught's Test). *Suppose the set $\Sigma \subseteq \text{Sent}(L)$ possesses only infinite models, and there exists a cardinal number $\kappa \geq \kappa_L$ such that every two models of $\Sigma$ of cardinality $\kappa$ are isomorphic. Then $\Sigma$ is complete.*

*Proof*: Let $\mathfrak{A}$ and $\mathfrak{B}$ be two models of $\Sigma$. By Corollary 2.1.2 there are $L$-structures $\mathfrak{A}_1$ and $\mathfrak{B}_1$ of cardinality $\kappa$ with $\mathfrak{A} \equiv \mathfrak{A}_1$ and $\mathfrak{B} \equiv \mathfrak{B}_1$. By hypothesis, $\mathfrak{A}_1 \cong \mathfrak{B}_1$. It then follows that

$$\mathfrak{A} \equiv \mathfrak{A}_1 \cong \mathfrak{B}_1 \equiv \mathfrak{B},$$

and thus $\mathfrak{A} \equiv \mathfrak{B}$. We now obtain the completeness of $\Sigma$ using Lemma 3.2.1. □

Now we would like to apply this test to three theories from §1.6.

**Theorem 3.2.3.** *The theories*
(1) *of dense linear orderings with extrema,*
(2) *of nontrivial, divisible, torsion-free, Abelian groups, and*
(3) *of algebraically closed fields of a fixed characteristic,*
*are complete.*

*Proof*: It is clear that each of these theories admits no finite model. The cardinality $\kappa_L$ of each language is $\aleph_0$.

(1) Let $\mathfrak{A} = \langle A; < \rangle$ and $\mathfrak{A}' = \langle A'; <' \rangle$ be two dense linear orderings without extrema. Furthermore, suppose $A$ and $A'$ are countable; let, say, $(b_n)_{n\in\mathbb{N}}$ be an enumeration of $A$, and $(b'_n)_{n\in\mathbb{N}}$ an enumeration of $A'$. With the help of Cantor's "back-and-forth procedure"[2] we define an isomorphism $\tau$ of $\mathfrak{A}$ and $\mathfrak{A}'$. We construct this isomorphism piece by piece, by defining new enumerations $(a_n)_{n\in\mathbb{N}}$ of $A$ and $(a'_n)_{n\in\mathbb{N}}$ of $A'$ in such a way that the mapping $\tau(a_n) := a'_n$ is automatically order-preserving.

We assume that we have already defined $a_n$ and $a'_n$ for $n < m$ such that

$$a_i < a_j \quad \text{iff} \quad a'_i < a'_j \tag{3.2.3.1}$$

for $i, j < m$. We must find $a_m$ and $a'_m$ such that the condition (3.2.3.1) continues to hold. Moreover, the choice of $a_m$ and $a'_m$ should be made in such a way that ultimately all $b_n$ and $b'_n$ are used. We do this as follows:

If $m$ is even, say $m = 2m_1$, then let $a_m$ be the $b_n$ with smallest index that does not lie in the set $\{a_0, \ldots, a_{m-1}\}$. Since $A'$ is a dense ordering without extrema, we can obviously choose $a'_m \in A'$ so that (3.2.3.1) holds for $i, j \leq m$.

If $m$ is odd, say $m = 2m_1 + 1$, then let $a'_m$ be the $b'_n$ with smallest index that does not lie in the set $\{a'_0, \ldots, a'_{m-1}\}$. By analogy with the previous case, we can choose $a_m \in A$ so that (3.2.3.1) continues to hold for $i, j \leq m$.

From the alternating character of the definitions, we see that for the above-mentioned minimal index $n$, $m_1 \leq n$ always holds. Therefore the mapping $\tau(a_m) = a'_m$ furnishes an isomorphism of $\mathfrak{A}$ and $\mathfrak{A}'$.

(2) Now let $\mathfrak{A}$ and $\mathfrak{A}'$ be divisible, torsion-free Abelian groups. Each such group is, in a canonical way, a vector space over the field $\mathbb{Q}$. Indeed, if $a \in |\mathfrak{A}|$ and $\frac{n}{m} \in \mathbb{Q}$, then we define $\frac{n}{m}a$ as follows: since $\mathfrak{A}$ is divisible, there is a $b \in |\mathfrak{A}|$ with $mb = a$. By torsion-freeness, $b$ is uniquely determined. Indeed, if $mb' = a$, then $m(b - b') = 0$ and thence $b = b'$ would follow. Thus we set $\frac{n}{m}a = nb$.

If we assume, moreover, that $\mathfrak{A}$ and $\mathfrak{A}'$ are of cardinality $\aleph_1$, then it follows that both $\mathfrak{A}$ and $\mathfrak{A}'$ have a basis over $\mathbb{Q}$ of cardinality $\aleph_1$. Therefore $\mathfrak{A}$ and $\mathfrak{A}'$ are isomorphic as $\mathbb{Q}$-vector spaces. Hence they are, in particular, isomorphic as Abelian groups.

(3) Finally, let $\mathfrak{A}$ and $\mathfrak{A}'$ be algebraically closed fields of equal characteristic, and let both be of cardinality $\aleph_1$. Then both possess transcendence bases of cardinality $\aleph_1$ over the prime field. Then by the well-known isomorphism theorem of Steinitz [Steinitz, 1930], $\mathfrak{A}$ and $\mathfrak{A}'$ are isomorphic. □

---

[2] This procedure was the inspiration for the proof of the Isomorphism Theorem 2.5.7.

In §§3.3 and 3.4 we shall treat the case of algebraically closed fields using other methods. The proof of the completeness of this theory in the case of fixed characteristic will then follow by means of algebraic facts much simpler than the isomorphism theorem of Steinitz.

This isomorphism theorem is, however, very important when considering a deep result of model theory, "Morley's theorem".

A theory $T$ (or an axiom system $\Sigma$) is called *$\kappa$-categorical* if every two models of $T$ (or $\Sigma$, respectively) of cardinality $\kappa$ are isomorphic.

We consider the case of a countable language $L$ (i.e. $\kappa_L = \aleph_0$) and an $L$-theory $T$ having no finite model. Then the completeness of $T$ follows from $\kappa$-categoricity, by Theorem 3.2.2. Morley's theorem says, now, that if such a theory is categorical in one uncountable cardinal number $\kappa$, then it is already categorical in every uncountable cardinal number. If follows that for such a theory $T$ there are only four possibilities:

(1) $T$ is categorical in every cardinality. An example of this is the theory of infinite sets. For the language of this theory, $I = J = K = \emptyset$; an axiom system is, say, $\Sigma = \{\alpha_{\geq n} \mid n \in \mathbb{N}\}$, where

$$\alpha_{\geq n} := \exists v_1, \ldots, v_n \left( \bigwedge_{i<j} v_i \neq v_j \right)$$

is a cardinality formula that says that there are at least $n$ distinct objects. The models of $\Sigma$ are, therefore, just the infinite sets. Two infinite sets of equal cardinality are naturally isomorphic, since an isomorphism in this language is nothing more than a bijection.

(2) $T$ is $\aleph_0$-categorical and not $\aleph_1$-categorical. This is the case for the theory of dense linear orderings without extrema, for example.

(3) $T$ is $\aleph_1$-categorical and not $\aleph_0$-categorical. The theory of (nontrivial) divisible, torsion-free, Abelian groups is an example of this, as is the theory of algebraically closed fields of a fixed characteristic.

(4) $T$ is neither $\aleph_0$-categorical nor $\aleph_1$-categorical. The theory of divisible, ordered, Abelian groups furnishes an example of this. This is actually the "general" case.

We cannot go into the proof of Morley's theorem in the framework of this book. The methods of that proof lead into a part of model theory that is currently in full bloom: stability theory. Compared to the part of model theory devoted more to the investigation of concrete, mostly algebraic theories, stability theory would better be called "abstract model theory". We refer the interested reader to the books of Shelah [1978] and Poizat [1985].

To conclude this section, we would like to derive several interesting consequences of the completeness of the theory of algebraically closed fields of a fixed characteristic. These involve a certain *transfer principle*.

**Theorem 3.2.4.** *If $\alpha$ is a sentence in the language of fields, then, writing "a.c." for "algebraically closed", we have:*

(1) *If $\alpha$ holds in an a.c. field, then $\alpha$ holds in every other a.c. field of the same characteristic.*

(2) *If $\alpha$ holds in all a.c. fields of prime characteristic, then $\alpha$ holds in all a.c. fields.*

(3) *If $\alpha$ holds in all a.c. fields of characteristic 0, then, for all prime numbers $p$ outside a finite set, $\alpha$ holds in all a.c. fields of prime characteristic $p$.*

(4) *If $\alpha$ is of the form $\forall x_1,\ldots,x_n \exists y_1,\ldots,y_m \delta$ with quantifier-free $\delta$, and if $\alpha$ holds in every finite field, then $\alpha$ holds in all algebraically closed fields.*

*Proof*: (1) This is nothing other than the completeness of the theory of a.c. fields of a fixed characteristic.

(2) By (1), it suffices to show that $\alpha$ holds in at least one a.c. field of characteristic 0, which means nothing other than the consistency of the following set $\Sigma$ of sentences:

$$\{\alpha\}\cup\{\mathrm{K}_1,\ldots,\mathrm{K}_9\}\cup\{\mathrm{AK}_n \mid n\geq 1\}\cup\{\neg\mathrm{C}_q \mid q\in\mathbb{N},\ q \text{ prime}\}.$$

Here, $\mathrm{K}_i, \mathrm{AK}_n, \mathrm{C}_q$ are the axioms introduced in (1.6.1.2), (1.6.1.3) and (1.6.1.4). A finite subset $\Pi$ of this axiom set $\Sigma$ can contain only finitely many characteristic sentences $\neg\mathrm{C}_q$; say, $\neg\mathrm{C}_{q_1},\ldots,\neg\mathrm{C}_{q_m}$. If $p$ is a prime number greater than $q_1,\ldots,q_m$, then obviously the algebraic closure $\widetilde{\mathbb{F}_p}$ of the Galois field $\mathbb{F}_p$ with $p$ elements is a model of $\Pi$. Since each finite subset of $\Sigma$ has a model, we obtain a model of $\Sigma$ itself, by the Finiteness Theorem 1.5.6.

(3) The hypothesis says that the set of sentences

$$\{\neg\alpha\}\cup\{\mathrm{K}_1,\ldots,\mathrm{K}_9\}\cup\{\mathrm{AK}_n \mid n\geq 1\}\cup\{\neg\mathrm{C}_q \mid q\in\mathbb{N},\ q \text{ prime}\}$$

has no model. By another use of Theorem 1.5.6, there is a finite subset $\Pi$ admitting no model. If $\neg\mathrm{C}_{q_1},\ldots,\neg\mathrm{C}_{q_m}$ are the characteristic sentences in $\Pi$, then it follows that every a.c. field $F$ satisfying those sentences necessarily violates $\neg\alpha$; i.e. if the characteristic of $F$ is different from $q_1,\ldots,q_m$, then $\alpha$ must hold in $F$.

(4) Let $p$ be a prime number. The algebraic closure $\widetilde{\mathbb{F}_p}$ of $\mathbb{F}_p$ is the union of its (uniquely determined) finite subfields $\mathbb{F}_{p^m}$ (with exactly $p^m$ elements). In particular,

$$\widetilde{\mathbb{F}_p} = \bigcup_{m\geq 1} \mathbb{F}_{p^{m!}},$$

where the finite fields $\mathbb{F}_{p^{m!}}$ now form an increasing chain. By hypothesis, $\alpha$ holds in every $\mathbb{F}_{p^{m!}}$, hence also in $\widetilde{\mathbb{F}_p}$, by Remark 2.4.7. The claim then follows from (1) and (2). □

From (4) it is easy to prove (following J. Ax) the following theorem of algebraic geometry:

**Theorem 3.2.5.** *Let $F$ be an algebraically closed field (e.g. the field $\mathbb{C}$ of complex numbers). Let $M$ be an algebraic subset of the affine space $F^n$. Further, let $f$ be a mapping from $M$ to itself with the following property: there are algebraic sets*

*$M_1,\dots,M_m$ with $M=\bigcup_{i=1}^m M_i$, and on each $M_i$, $f$ is, coordinatewise, a polynomial; i.e. there are $p_{ij}\in F[X_1,\dots,X_n]$ with*

$$f(a_1,\dots,a_n)=(p_{i1}(a_1,\dots,a_n),\dots,p_{in}(a_1,\dots,a_n)),$$

*for all $(a_1,\dots,a_n)\in M_i$. Then if $f$ is injective, then $f$ is also surjective.*

*Proof*: If we can write the conclusion of this theorem as an $\forall\exists$-sentence $\alpha$, then, by Theorem 3.2.4(4), the theorem will have been proved, since $\alpha$ holds in every finite field. (The set $M$ is, then, likewise finite, and an injection of $M$ into itself is naturally also surjective.)

First, we consider polynomials $q_{i1},\dots,q_{ir}\in F[X_1,\dots,X_n]=:F\left[\overline{X}\right]$ that define the set $M_i$ (for $1\le i\le m$). Here we can take the number $r$ to be the same for all $M_i$. (This is always achievable by repetition of polynomials.) Further, let $f$ be defined by $p_{i1},\dots,p_{in}\in F\left[\overline{X}\right]$ on $M_i$. Now we must formulate the conclusion of this theorem. For simplicity, we consider only the case $m=1$, and leave the general case to the reader. We therefore assume that $f$ is defined on $M$ by the polynomials $p_1,\dots,p_n$, and $M$ itself is defined by $q_1,\dots,q_r$. Then we must formulate:

$$\text{“}f\text{ goes from }M\text{ to }M\text{”}\wedge\text{“}f\text{ injective”}\;\rightarrow\;\text{“}f\text{ surjective”}. \tag{3.2.5.1}$$

The first conjunct in the premise becomes:

$$\forall\overline{x}\left(\bigwedge_{j=1}^{r}q_j(\overline{x})\doteq 0\;\rightarrow\;\bigwedge_{j=1}^{r}q_j(p_1(\overline{x}),\dots,p_n(\overline{x}))\doteq 0\right).$$

The second conjunct in the premise becomes:

$$\forall\overline{x}\,\forall\overline{y}\left(\bigwedge_{j=1}^{r}(q_j(\overline{x})\doteq 0\wedge q_j(\overline{y})\doteq 0)\wedge\bigwedge_{i=1}^{n}p_i(\overline{x})\doteq p_i(\overline{y})\;\rightarrow\;\bigwedge_{i=1}^{n}x_i\doteq y_i\right).$$

The conclusion of (3.2.5.1) becomes:

$$\forall\overline{u}\,\exists\overline{w}\left(\bigwedge_{j=1}^{r}(q_j(\overline{u})\doteq 0\;\rightarrow\;\bigwedge_{j=1}^{r}q_j(\overline{w})\doteq 0\wedge\bigwedge_{i=1}^{n}u_i\doteq p_i(\overline{w})\right).$$

Altogether, we therefore have a sentence of the form

$$\forall\overline{x}\,\delta_1\;\wedge\;\forall\overline{x}\,\forall\overline{y}\,\delta_2\;\rightarrow\;\forall\overline{u}\,\exists\overline{w}\,\delta_3, \tag{3.2.5.2}$$

where $\delta_1,\delta_2,\delta_3$ are quantifier-free. By pure logic, (3.2.5.2) is equivalent to

$$\forall\overline{u}\,\exists\overline{w}\,\exists\overline{x}\,\exists\overline{y}\,(\delta_1\wedge\delta_2\rightarrow\delta_3);$$

but this is an $\forall\exists$-sentence.

So far, we have not taken the coefficients of the polynomials $p_i$ and $q_j$ at all into account. This gap is remedied as follows. We replace the coefficients by variables $v_k$ that have not been used up till now. (Observe that the $\overline{u}, \overline{w}, \overline{x}, \overline{y}$ are, in reality, also certain variables $v_\mu$.) The number of these variables depends on the degrees of the polynomials (and on $r$ and $n$). Denote the result of replacing the coefficients in $\delta_1, \delta_2, \delta_3$ by $\delta_1', \delta_2', \delta_3'$.

Then we see, finally, that the $\forall\exists$-sentence

$$\forall\overline{v}\, \forall\overline{u}\, \exists\overline{w}\, \exists\overline{x}\, \exists\overline{y}\, (\delta_1' \wedge \delta_2' \to \delta_3')$$

expresses the conclusion of the theorem (with $m = 1$) – at least for polynomials $p_i$ and $q_j$ up to a certain degree. However, since such a sentence can be constructed for every degree, the theorem is thus proved. □

## 3.3 Model Completeness

In this section we shall introduce a property of theories – model completeness – with whose help we will be able to prove the completeness of the three remaining theories whose completeness we had claimed in §1.6. But we shall not do this, since their completeness can be proved more easily in the next section with an even stronger property of theories: quantifier elimination. On the whole, however, the concept of model completeness, and even more so the related concept of a structure that is existentially closed in a class of structures, is of eminent significance, especially with regard to possible applications.

A set $\Sigma \subseteq \mathrm{Sent}(L)$ is called *model complete* if for each model $\mathfrak{A}$ of $\Sigma$, the set of $L(\mathfrak{A})$-sentences $\Sigma \cup D(\mathfrak{A})$ is complete. Recall that $D(\mathfrak{A})$ is the diagram of $\mathfrak{A}$, defined just before (2.4.2).

**Lemma 3.3.1.** *$\Sigma \subseteq \mathrm{Sent}(L)$ is model complete if and only if, for any two models $\mathfrak{A}, \mathfrak{B}$ of $\Sigma$ with $\mathfrak{A} \subseteq \mathfrak{B}$, we even have $\mathfrak{A} \preceq \mathfrak{B}$.*

*Proof*: First suppose $\Sigma$ is model complete. If $\mathfrak{A}$ and $\mathfrak{B}$ are models of $\Sigma$ with $\mathfrak{A} \subseteq \mathfrak{B}$, then obviously the $L(\mathfrak{A})$-structures $(\mathfrak{A}, |\mathfrak{A}|)$ and $(\mathfrak{B}, |\mathfrak{A}|)$ are both models of $\Sigma \cup D(\mathfrak{A})$. We therefore have

$$(\mathfrak{A}, |\mathfrak{A}|) \equiv (\mathfrak{B}, |\mathfrak{A}|),$$

by Lemma 3.2.1. This implies $\mathfrak{A} \preceq \mathfrak{B}$, by Corollary 2.4.4.

Now suppose, conversely, that $\mathfrak{A} \preceq \mathfrak{B}$ whenever $\mathfrak{A}$ and $\mathfrak{B}$ are models of $\Sigma$ with $\mathfrak{A} \subseteq \mathfrak{B}$. Then we fix a model $\mathfrak{A}$ of $\Sigma$, and show the completeness of $\Sigma \cup D(\mathfrak{A})$. Obviously, $(\mathfrak{A}, |\mathfrak{A}|)$ is a model of it. By Lemma 3.2.1, it will suffice to show that every other model of $\Sigma \cup D(\mathfrak{A})$ is elementarily equivalent to $(\mathfrak{A}, |\mathfrak{A}|)$. By Lemma 2.4.2, such a model contains an isomorphic copy of $\mathfrak{A}$. If we identify $\mathfrak{A}$ with its image, we may assume that the model being considered is of the form $(\mathfrak{B}, |\mathfrak{A}|)$, where even

$\mathfrak{A} \subseteq \mathfrak{B}$ holds. $\mathfrak{B}$ is, like $\mathfrak{A}$, a model of $\Sigma$. Therefore $\mathfrak{A} \preceq \mathfrak{B}$ holds, by hypothesis. But then

$$(\mathfrak{A}, |\mathfrak{A}|) \equiv (\mathfrak{B}, |\mathfrak{A}|)$$

follows, by Corollary 2.4.4. □

A theory that is model complete need not be complete. For example, as we shall see in Theorem 3.3.4, the theory of algebraically closed fields (of arbitrary characteristic) is model complete, but not complete, since the characteristic of a field can be expressed by means of the sentences $C_p$ (recall (1.6.1.4)).

However, if a model complete axiom system $\Sigma \subseteq \mathrm{Sent}(L)$ admits a prime model, then we can deduce the completeness of $\Sigma$. Here, an $L$-structure $\mathfrak{P}$ is called a *prime model* of $\Sigma$ if $\mathfrak{P}$ is a model of $\Sigma$ and $\mathfrak{P}$ can be embedded isomorphically in every other model of $\Sigma$.

**Corollary 3.3.2.** *Suppose the set $\Sigma \subseteq \mathrm{Sent}(L)$ is model complete and admits a prime model $\mathfrak{P}$. Then $\Sigma$ is complete.*

*Proof*: Let $\mathfrak{A}_1$ and $\mathfrak{A}_2$ be models of $\Sigma$. We identify $\mathfrak{P}$ with its isomorphic images in $\mathfrak{A}_1$ and $\mathfrak{A}_2$. Then $\mathfrak{P} \subseteq \mathfrak{A}_i$, whence $\mathfrak{P} \preceq \mathfrak{A}_i$ $(i = 1,2)$, immediately, by model completeness. In particular, $\mathfrak{A}_1 \equiv \mathfrak{P} \equiv \mathfrak{A}_2$. Then $\Sigma$ is complete, by Lemma 3.2.1. □

Now we wish to derive a very useful criterion for proving the model completeness of a theory. For this, we need several definitions. We shall call an $L$-formula *existential*, or an *$\exists$-formula*, if it is of the form

$$\exists x_1, \ldots, x_n\ \psi,$$

with $\psi$ quantifier-free (this generalizes the definition in Theorem 2.5.4). Analogously, we call a formula of the form

$$\forall x_1, \ldots, x_n\ \psi,$$

with $\psi$ quantifier-free, *universal*, or an *$\forall$-formula*. If $\mathfrak{A}$ and $\mathfrak{B}$ are $L$-structures with $\mathfrak{A} \subseteq \mathfrak{B}$, then in Corollary 2.5.5 we called $\mathfrak{A}$ *existentially closed* in $\mathfrak{B}$ if each $\exists$-sentence of the language $L(\mathfrak{A})$ that holds in $\mathfrak{B}$ also holds in $\mathfrak{A}$. This is equivalent, by Lemma 2.4.1, to the condition that for each $\exists$-formula $\varphi$ of the language $L$, and for each evaluation $h$ in $\mathfrak{A}$,

$$\mathfrak{B} \models \varphi[h] \quad \text{implies} \quad \mathfrak{A} \models \varphi[h].$$

**Theorem 3.3.3** (Robinson's Test). *For $\Sigma \subseteq \mathrm{Sent}(L)$, the following are equivalent:*

(1) *$\Sigma$ is model complete,*

(2) *for every two models $\mathfrak{A}, \mathfrak{B}$ of $\Sigma$ with $\mathfrak{A} \subseteq \mathfrak{B}$, $\mathfrak{A}$ is existentially closed in $\mathfrak{B}$,*

(3) *to each $L$-formula $\varphi$ there is a universal $L$-formula $\rho$ with $\mathrm{Fr}(\rho) \subseteq \mathrm{Fr}(\varphi)$ such that $\Sigma \vdash \forall(\varphi \leftrightarrow \rho)$.*

*Proof*: (1) $\Rightarrow$ (2) follows immediately from Lemma 3.3.1.

(2) $\Rightarrow$ (3): First let $\varphi$ be an existential $L$-formula with $\mathrm{Fr}(\varphi) \subseteq \{v_0, \ldots, v_n\}$. We extend the language $L$ with new constant symbols $c_0, \ldots, c_n$ (without loss of generality we may assume $\{0, \ldots, n\} \cap K = \emptyset$). Let

$$\varphi' = \varphi(v_0/c_0, \ldots, v_n/c_n).$$

Let the extended language be called $L'$.

Let $\mathfrak{A}', \mathfrak{B}'$ be two $L'$-models of $\Sigma$ with $\mathfrak{A}' \subseteq \mathfrak{B}'$. For the $L$-restrictions $\mathfrak{A}$ and $\mathfrak{B}$ of $\mathfrak{A}'$ and $\mathfrak{B}'$, respectively, it follows, by (2), that $\mathfrak{A}$ is existentially closed in $\mathfrak{B}$. From this we get, by Lemma 2.4.1:

$$\mathfrak{B}' \models \varphi' \quad \text{implies} \quad \mathfrak{A}' \models \varphi'.$$

Therefore, by Theorem 3.1.7, there is a quantifier-free $L'$-formula $\delta'$ with

$$\Sigma \vdash (\varphi' \leftrightarrow \forall \delta').$$

Here we can, if necessary, arrange, by renaming the (free) variables occurring in $\delta'$, that $\mathrm{Fr}(\delta') \cap \{v_0, \ldots, v_n\} = \emptyset$. Then, if we set

$$\delta = \delta'(c_0/v_0, \ldots, c_n/v_n),$$

we obtain, using Lemma 1.4.3, that

$$\Sigma \vdash \forall(\varphi \leftrightarrow \forall x_1, \ldots, x_m\, \delta),$$

in case $\mathrm{Fr}(\delta') = \{x_1, \ldots, x_m\}$. Hence we have proved that, "modulo $\Sigma$", each existential $L$-formula is equivalent to a universal $L$-formula. Now we obtain the claim (3) by induction on the recursive construction of an arbitrary $L$-formula $\varphi$:

If $\varphi$ is an atomic formula, then we set $\rho = \varphi$.

Suppose $\varphi = \neg\varphi_1$. By the inductive hypothesis, there is a universal formula $\rho_1$ such that $\mathrm{Fr}(\rho_1) \subseteq \mathrm{Fr}(\varphi_1)$ and

$$\Sigma \vdash (\varphi_1 \leftrightarrow \rho_1).$$

Then we have, naturally,

$$\Sigma \vdash (\neg\varphi_1 \leftrightarrow \neg\rho_1).$$

As we have just seen, $\neg\rho_1$ is, modulo $\Sigma$, equivalent to a universal $L$-formula $\rho$ with $\mathrm{Fr}(\rho) \subseteq \mathrm{Fr}(\neg\rho_1)$. We therefore obtain

$$\Sigma \vdash (\neg\varphi_1 \leftrightarrow \rho)$$

with $\mathrm{Fr}(\rho) \subseteq \mathrm{Fr}(\neg\varphi_1)$.

Next, if $\varphi = (\varphi_1 \vee \varphi_2)$, then there are, by the inductive hypothesis, universal formulae $\rho_i$, for $i = 1, 2$, such that $\mathrm{Fr}(\rho_i) \subseteq \mathrm{Fr}(\varphi_i)$ and

$$\Sigma \vdash (\varphi_i \leftrightarrow \rho_i).$$

Then

$$\Sigma \vdash ((\varphi_1 \wedge \varphi_2) \leftrightarrow (\rho_1 \wedge \rho_2)).$$

By pure logic, $(\rho_1 \wedge \rho_2)$ is again equivalent to a universal formula $\rho$ with $\mathrm{Fr}(\rho) \subseteq \mathrm{Fr}(\rho_1 \wedge \rho_2)$. We therefore obtain, finally,

$$\Sigma \vdash (\varphi \leftrightarrow \rho) \quad \text{with} \quad \mathrm{Fr}(\rho) \subseteq \mathrm{Fr}(\varphi).$$

Finally, let $\varphi = \forall x \varphi_1$. Then by the inductive hypothesis, there is a universal formula $\rho_1$ such that $\mathrm{Fr}(\rho_1) \subseteq \mathrm{Fr}(\varphi_1)$ and

$$\Sigma \vdash (\varphi_1 \leftrightarrow \rho_1).$$

Then we immediately deduce

$$\Sigma \vdash (\forall x \varphi_1 \leftrightarrow \forall x \rho_1)$$

and also $\mathrm{Fr}(\forall x \rho_1) \subseteq \mathrm{Fr}(\forall x \varphi_1)$. Condition (3) is thus proved.

(3) $\Rightarrow$ (1): We apply Lemma 3.3.1. Let $\mathfrak{A}$ and $\mathfrak{B}$ be models of $\Sigma$ with $\mathfrak{A} \subseteq \mathfrak{B}$. Further, let $\varphi$ be an $L$-formula and $h$ an evaluation in $\mathfrak{A}$. By (3), there is a universal $L$-formula $\rho$ with $\Sigma \vdash (\varphi \leftrightarrow \rho)$. Then from $\mathfrak{B} \models \varphi[h]$ follows $\mathfrak{B} \models \rho[h]$, naturally. As an immediate consequence, $\mathfrak{A} \models \rho[h]$, by the universality of $\rho$. This is equivalent to $\mathfrak{A} \models \varphi[h]$. We have thus proved $\mathfrak{A} \preceq \mathfrak{B}$. □

With the help of Robinson's Test we can see immediately that, e.g. the theory of linear orderings without extrema is model complete. For this, let $\mathfrak{A} = \langle A; < \rangle$ and $\mathfrak{A}' = \langle A'; <' \rangle$ be models of this theory with $\mathfrak{A} \subseteq \mathfrak{A}'$. Further, let $\exists x_1, \ldots, x_n \delta$ be an $L(A)$-sentence with quantifier-free $\delta$. Let the constant symbols $\underline{a}$ occurring in $\delta$ be $\underline{a_1}, \ldots, \underline{a_m}$, for certain $a_1, \ldots, a_m \in A$. Then

$$(\mathfrak{A}', A) \models \exists x_1, \ldots, x_n \delta$$

claims the existence of certain elements $b_1, \ldots, b_n$ in $A'$ occupying "certain locations" in relation to the $a_1, \ldots, a_m$ and to each other. Indeed, the formula $\delta$ can express nothing more (for this, recall also Theorem 3.2.3(1)). Now, since $\langle A; < \rangle$ is dense and without extrema, we see immediately that we can find such elements $b_1, \ldots, b_n$ in $A$; i.e. we see that

$$(\mathfrak{A}, A) \models \exists x_1, \ldots, x_n \delta.$$

By Theorem 3.3.3, the model completeness of the theory of dense linear orderings without extrema is thus proved.

If we try to proceed analogously with the theory of algebraically closed fields, we run into difficulties. Here it is advisable to work with Corollary 2.5.5, which presents a very practical method for proving existential closedness.

**Theorem 3.3.4.** *The theory of algebraically closed fields is model complete.*

*Proof*: Let $\mathfrak{A}$ and $\mathfrak{B}$ be algebraically closed fields with $\mathfrak{A} \subseteq \mathfrak{B}$. We choose a $\kappa$-saturated, elementary extension $\mathfrak{A}'$ of $\mathfrak{A}$ (where we may assume that $\kappa > \text{card}(|\mathfrak{B}|)$), by Theorem 2.5.2. From this follows, first, that $\mathfrak{A}'$ is also an algebraically closed field, and, second, that $\text{card}(|\mathfrak{A}'|) \geq \kappa > \text{card}(|\mathfrak{A}|)$. In particular, $\mathfrak{A}'$ then has infinite transcendence degree over $\mathfrak{A}$. Therefore every subfield of $\mathfrak{B}$ that is finitely generated over $\mathfrak{A}$ is embeddable in $\mathfrak{A}'$, by a theorem of Steinitz.

Then by Corollary 2.5.5(1)(b) and (2), $\mathfrak{A}$ is existentially closed in $\mathfrak{B}$. □

From the model completeness of the theory of algebraically closed fields, we deduce immediately, using Corollary 3.3.2, the completeness in the case of fixed characteristic. Namely, if the characteristic is 0, then the algebraic closure of $\mathbb{Q}$ is a prime model. And if the characteristic is a prime number $p$, then the algebraic closure of $\mathbb{F}_p$ is a prime model.

As a small application of Theorem 3.3.4, we prove the "Hilbert Nullstellensatz", which actually asserts nothing other than the model completeness of the theory of algebraically closed fields.

**Theorem 3.3.5** (Hilbert Nullstellensatz). *Let $I$ be a proper ideal in the polynomial ring $F[X_1,\ldots,X_n]$ over an algebraically closed field $F$. Then there exist $a_1,\ldots,a_n \in F$ with $f(a_1,\ldots,a_n) = 0$ for all $f \in I$.*

*Proof*: Let $M$ be a maximal ideal containing $I$. The existence of such an ideal $M$ is guaranteed by Zorn's Lemma. By the Hilbert Basis Theorem, $M$ is generated by finitely many polynomials $f_1,\ldots,f_m \in F[X_1,\ldots,X_n]$. Since $M$ is maximal, the residue ring

$$F_1 = F[X_1,\ldots,X_n] \,/\, M$$

is again a field – which can, obviously, be regarded as an extension field of $F$. The polynomials $f_1,\ldots,f_m$ possess a common zero in $F_1$, namely, $(X_1+M,\ldots,X_n+M)$, whose coordinates are the residue classes of $X_1,\ldots,X_n$ with respect to the ideal $M$. Let $\widetilde{F_1}$ be the algebraic closure of $F_1$. Then we obtain, clearly,

$$(\widetilde{F_1},|F|) \models \exists x_1,\ldots,x_n\,\big(\underline{f_1}(x_1,\ldots,x_n) \doteq 0 \wedge \cdots \wedge \underline{f_m}(x_1,\ldots,x_n) \doteq 0\big).$$

Here, $\underline{f_i}$ denotes the $L(F)$-term that one obtains when one replaces the coefficients $b_j$ in the polynomials $f_i$ by the constant symbols $\underline{b_j}$. By Theorem 3.3.4, $F$ is an elementary substructure of $\widetilde{F_1}$, which implies

$$(F,|F|) \models \exists x_1,\ldots,x_n\,\big(\underline{f_1}(x_1,\ldots,x_n) \doteq 0 \wedge \cdots \wedge \underline{f_m}(x_1,\ldots,x_n) \doteq 0\big);$$

i.e. there exist $a_1,\ldots,a_n \in F$ with $f_j(a_1,\ldots,a_n) = 0$ for $1 \leq j \leq m$. However, since $M$ is generated by $f_1,\ldots,f_m$, we even have $f(a_1,\ldots,a_n) = 0$ for all $f \in M$, and, in particular, for all $f \in I$. □

If $\Sigma$ is model complete, then Theorem 2.4.6 yields a remarkable property of $\text{Mod}(\Sigma)$. Namely, if $(\mathfrak{A}_\nu)_{\nu<\alpha}$ is an $\alpha$-chain of models of $\Sigma$, then this is already an elementary chain, due to the model completeness of $\Sigma$. Therefore the union $\bigcup_{\nu<\alpha}\mathfrak{A}_\nu$ is, by Theorem 2.4.6, itself a model of $\Sigma$. (Note that for limit ordinals $\lambda<\alpha$, $\bigcup_{\nu<\lambda}\mathfrak{A}_\nu=\mathfrak{A}_\lambda$ need not hold. However, by model completeness, we must have $\bigcup_{\nu<\lambda}\mathfrak{A}_\nu\preceq\mathfrak{A}_\lambda$. We may therefore apply Theorem 2.4.6 to the chain $(\mathfrak{A}'_\nu)_{\nu<\alpha}$ defined by $\mathfrak{A}'_{\nu+1}=\mathfrak{A}_\nu$ and $\mathfrak{A}'_\nu=\bigcup_{\nu<\lambda}\mathfrak{A}_\nu$ for limit ordinals $\lambda<\alpha$.)

We shall call a class $\mathscr{M}$ of $L$-structures *inductive* if it contains the union $\bigcup_{\nu<\alpha}\mathfrak{A}_\nu$ of every $\alpha$-chain $(\mathfrak{A}_\nu)_{\nu<\alpha}$ of structures in $\mathscr{M}$. If such an $\mathscr{M}$ equals $\mathrm{Mod}(\Sigma)$, then we call $\Sigma$ *inductive*. A model complete axiom system $\Sigma$ is, therefore, always inductive. As we shall see, this has consequences for possible alternate axiomatizations of $\mathrm{Ded}(\Sigma)$. First, however, we need several preparatory results.

For a set $\Sigma$ of sentences, we define

$$\Sigma_{\forall\exists}=\{\,\alpha\mid\Sigma\vdash\alpha,\text{ and }\alpha\text{ is an }\forall\exists\text{-sentence}\,\}.$$

First, we prove:

**Lemma 3.3.6.** *Each model $\mathfrak{A}$ of $\Sigma_{\forall\exists}$ is isomorphic to an existentially closed substructure of a model $\mathfrak{B}$ of $\Sigma$.*

*Proof*: We consider the set

$$\Sigma'=\Sigma\cup\mathrm{Th}(\mathfrak{A},A)_\forall$$

of $L(\mathfrak{A})$-sentences, where $A=|\mathfrak{A}|$ and

$$\mathrm{Th}(\mathfrak{A},A)_\forall=\{\,\alpha\in L(\mathfrak{A})\mid(\mathfrak{A},A)\models\alpha,\text{ and }\alpha\text{ is universal}\,\}.$$

First we show that $\Sigma'$ admits a model; for this it will suffice, according to the Compactness Theorem, to find a model for every finite subset thereof, and it will therefore suffice to find a model for every set

$$\Pi':=\Sigma\cup\{\alpha_1,\ldots,\alpha_m\},$$

with $\alpha_1,\ldots,\alpha_m\in\mathrm{Th}(\mathfrak{A},A)_\forall$. Since $\alpha_1,\ldots,\alpha_m$ are universal, so is $(\alpha_1\wedge\cdots\wedge\alpha_m)$ (up to logical equivalence). We may therefore restrict to the case

$$\Pi':=\Sigma\cup\{\forall x_1,\ldots,x_n\,\delta(x_1,\ldots,x_n,\underline{a_1},\ldots,\underline{a_r})\},$$

where $\delta\in\mathrm{Fml}(L)$ is quantifier-free with variables $x_1,\ldots,x_n$ and $y_1,\ldots,y_r$, and where

$$(\mathfrak{A},A)\models\forall x_1,\ldots,x_n\,\delta(x_1,\ldots,x_n,\underline{a_1},\ldots,\underline{a_r}).$$

Here the variables $y_1,\ldots,y_r$ are replaced by the constants $\underline{a_1},\ldots,\underline{a_r}$ for certain elements $a_1,\ldots,a_r\in A$.

Because of this property and the hypothesis $\mathfrak{A}\models\Sigma_{\forall\exists}$, the sentence

$$\neg\exists y_1,\ldots,y_r\,\forall x_1,\ldots,x_n\,\delta(x_1,\ldots,x_n,y_1,\ldots,y_r)$$

obviously does not belong to $\Sigma_{\forall\exists}$; i.e. it does not hold in all models of $\Sigma$. Thus, there is a model $\mathfrak{C}$ of $\Sigma$ with

$$\mathfrak{C} \models \exists y_1,\ldots,y_r \forall x_1,\ldots,x_n \,\delta(x_1,\ldots,x_n,y_1,\ldots,y_r).$$

For a suitable interpretation of $\underline{a_1},\ldots,\underline{a_n}$, we have thus found a model of $\Pi'$.

Now let $\mathfrak{B}'$ be a model of $\overline{\Sigma'}$. Since $\mathfrak{B}'$ is, in particular, a model of $D(\mathfrak{A})$ (this is contained in $\mathrm{Th}(\mathfrak{A},A)_\forall$), we can, using Lemma 2.4.2, assume that $\mathfrak{B}'$ contains $\mathfrak{A}$. We therefore have $\mathfrak{B}' = (\mathfrak{B},A)$, where the $L$-structure $\mathfrak{B}$ is, then, a model of $\Sigma$. It still remains to show that $\mathfrak{A}$ is existentially closed in $\mathfrak{B}$. But this is clear: namely, if $\alpha$ is a universal $L(\mathfrak{A})$-formula with $(\mathfrak{A},A) \models \alpha$, then $\alpha$ lies in $\mathrm{Th}(\mathfrak{A},A)_\forall$. But then $\alpha$ holds also in $(\mathfrak{B},A)$. This is clearly equivalent to the existential closedness of $\mathfrak{A}$ in $\mathfrak{B}$. □

**Theorem 3.3.7.** *For* $\Sigma \subseteq \mathrm{Sent}(L)$, $\Sigma$ *is inductive if and only if* $\mathrm{Mod}(\Sigma) = \mathrm{Mod}(\Sigma_{\forall\exists})$. *In particular,* $\mathrm{Mod}(\Sigma) = \mathrm{Mod}(\Sigma_{\forall\exists})$ *in case* $\Sigma$ *is model complete.*

*Proof*: Trivially,

$$\mathrm{Mod}(\Sigma) \subseteq \mathrm{Mod}(\Sigma_{\forall\exists})$$

always holds. Now we assume that $\mathrm{Mod}(\Sigma)$ is inductive, and fix a $\mathfrak{A}_0 \in \mathrm{Mod}(\Sigma_{\forall\exists})$. By Lemma 3.3.6, there is an extension $\mathfrak{B}_0$ of $\mathfrak{A}_0$ that is a model of $\Sigma$ and in which $\mathfrak{A}_0$ is existentially closed. By Theorem 2.5.2 there is an elementary, $\kappa$-saturated extension $\mathfrak{A}_1$ of $\mathfrak{A}_0$, where $\kappa$ may be chosen $> \mathrm{card}(|\mathfrak{B}_0|)$. (If $\mathfrak{A}_0$ is finite, then $\mathfrak{A}_0$ itself is $\kappa$-saturated.) Since $\mathfrak{A}_0$ is existentially closed in $\mathfrak{B}_0$, the structure $\mathfrak{B}_0$ can, by Corollary 2.5.5(1)(a), be identified with a substructure of $\mathfrak{A}_1$. In view of $\mathfrak{A}_0 \preceq \mathfrak{A}_1$, $\mathfrak{A}_1$ is again a model of $\Sigma_{\forall\exists}$, and we can repeat the above-described process.

After infinitely many steps, we thereby obtain a chain

$$\mathfrak{A}_0 \subseteq \mathfrak{B}_0 \subseteq \mathfrak{A}_1 \subseteq \mathfrak{B}_1 \subseteq \cdots \subseteq \mathfrak{A}_n \subseteq \mathfrak{B}_n \subseteq \cdots,$$

where the subchain

$$\mathfrak{A}_0 \subseteq \mathfrak{A}_1 \subseteq \mathfrak{A}_2 \cdots \subseteq \mathfrak{A}_n \subseteq \cdots$$

is elementary. Then Theorem 2.4.6 implies

$$\mathfrak{A}_0 \preceq \bigcup_{n\in\mathbb{N}} \mathfrak{A}_n.$$

Moreover, all the $\mathfrak{B}_n$ are models of $\Sigma$, which, by hypothesis, has

$$\bigcup_{n\in\mathbb{N}} \mathfrak{B}_n \models \Sigma$$

as a consequence. If we now consider the obvious equation

$$\bigcup_{n\in\mathbb{N}} \mathfrak{A}_n = \bigcup_{n\in\mathbb{N}} \mathfrak{B}_n,$$

then we see, finally, that $\mathfrak{A}_0$ is a model of $\Sigma$. Thus we have proved the relation $\mathrm{Mod}(\Sigma) = \mathrm{Mod}(\Sigma_{\forall\exists})$ for inductive $\Sigma$.

The converse follows immediately from Remark 2.4.7. □

Again let $\Sigma \subseteq \mathrm{Sent}(L)$. We shall call an axiom system $\Sigma^* \subseteq \mathrm{Sent}(L)$ a (*strict*) *model companion*[3] of $\Sigma$ if:

(i) every model of $\Sigma^*$ is also a model of $\Sigma$.
(ii) every model of $\Sigma$ can be extended to a model of $\Sigma^*$.
(iii) $\Sigma^*$ is model complete.

Instead of considering an axiom system $\Sigma$, we can consider the theory $T = \mathrm{Ded}(\Sigma)$ belonging to it. Then (i) says precisely that $T^* = \mathrm{Ded}(\Sigma^*)$ is an extension theory of $T$.

As examples, we have, so far, (a) the theory $T^*$ of dense linear orderings without extrema as the model companion of, say, the theory $T$ of linear orderings (axiomatized by $O_1$, $O_2$, $O_3$ of §1.6), and (b) the theory $T^*$ of algebraically closed fields as the model companion of, say, the theory $T$ of fields (axiomatized by $K_1$–$K_9$ of §1.6). In the first case, we observe that for a linear ordering $\mathfrak{A} = \langle A; < \rangle$, the lexicographically ordered product $A \times \mathbb{Q}$ is a model of $T^*$. In the second case, the algebraic closure of a field is a model of $T^*$.

In general, a theory $T$ need not admit a model companion – e.g. the theory of commutative rings has no model companion. However, if there is a model companion $T^*$ to a theory $T$, then it is uniquely determined. This, and the manner in which the model class of $T^*$ can be determined, are stated in the following theorem; to formulate it, we need one more definition.

If $\mathcal{M}$ is a class of $L$-structures, then we call $\mathfrak{A} \in \mathcal{M}$ *existentially closed in* $\mathcal{M}$ if $\mathfrak{A}$ is existentially closed in each $\mathfrak{B} \in \mathcal{M}$ with $\mathfrak{A} \subseteq \mathfrak{B}$. We set

$$E(\mathcal{M}) = \{\mathfrak{A} \in \mathcal{M} \mid \mathfrak{A} \text{ is existentially closed in } \mathcal{M}\}.$$

If $\mathcal{M} = \mathrm{Mod}(\Sigma)$, then we write $E(\Sigma)$ for $E(\mathrm{Mod}(\Sigma))$.

**Theorem 3.3.8.** *Let $T$ and $T^*$ be $L$-theories.*

(1) *If $T^*$ is a model companion of $T$, then* $\mathrm{Mod}(T^*) = E(T)$.
(2) *If $T$ is inductive and $E(T)$ is axiomatizable – say, $E(T) = \mathrm{Mod}(\Sigma^*)$ –, then $T^* = \mathrm{Ded}(\Sigma^*)$ is a model companion of $T$.*

*Proof*: (1) If $\mathfrak{A}$ is a model of $T^*$ and $\mathfrak{A} \subseteq \mathfrak{B}$, where $\mathfrak{B}$ is a model of $T$, then we find, by (ii), a model $\mathfrak{C}$ of $T^*$ with

$$\mathfrak{A} \subseteq \mathfrak{B} \subseteq \mathfrak{C}.$$

By (iii), $T^*$ is model complete, whence $\mathfrak{A} \preceq \mathfrak{C}$. From this it follows that, for every existential $L(\mathfrak{A})$-sentence $\delta$, if $(\mathfrak{B}, |\mathfrak{A}|) \models \delta$, then, first, $(\mathfrak{C}, |\mathfrak{A}|) \models \delta$, and, finally, $(\mathfrak{A}, |\mathfrak{A}|) \models \delta$.

[3] The reason for including the word "strict" will become clear in Exercise 3.5.5.

We have hence proved

$$\mathrm{Mod}(T^*) \subseteq E(T).$$

For the opposite inclusion, we consider an $\mathfrak{A} \in E(T)$, and extend it using (ii) to a model $\mathfrak{B}$ of $T^*$. Since $\mathfrak{A}$ is existentially closed in $\mathfrak{B}$, we can, using Theorem 2.5.2 and Corollary 2.5.5(1)(a), embed the structure $\mathfrak{B}$ in a sufficiently saturated elementary extension $\mathfrak{C}$ of $\mathfrak{A}$. We therefore have the following situation:

$$\mathfrak{A} \subseteq \mathfrak{B} \subseteq \mathfrak{C} \quad \text{and} \quad \mathfrak{A} \preceq \mathfrak{C}.$$

By Theorem 3.3.7, $\mathrm{Mod}(T^*) = \mathrm{Mod}(T^*_{\forall\exists})$. It therefore suffices to show that every $\forall\exists$-sentence $\varphi$ in $T^*$ holds in $\mathfrak{A}$.

Let $\varphi \in T^*$ be of the form

$$\varphi = \forall x_1,\ldots,x_n \exists y_1,\ldots,y_m\, \delta,$$

with quantifier-free $\delta$. Since $\mathfrak{B} \models T^*$, we have, in particular, $\mathfrak{B} \models \varphi$. Now if $a_1,\ldots,a_n \in |\mathfrak{A}| = A$, then we have

$$(\mathfrak{B},A) \models \exists y_1,\ldots,y_m\, \delta\big(x_1/\underline{a_1},\ldots,x_n/\underline{a_n}\big).$$

This transfers, trivially, to $(\mathfrak{C},A)$, and thence, finally, using $\mathfrak{A} \preceq \mathfrak{C}$, to $(\mathfrak{A},A)$. Altogether, we thus obtain $\mathfrak{A} \models \varphi$.

(2) If $E(T) = \mathrm{Mod}(\Sigma^*)$, then, in particular, for $\mathfrak{A}, \mathfrak{B} \in \mathrm{Mod}(\Sigma^*)$ with $\mathfrak{A} \subseteq \mathfrak{B}$, $\mathfrak{A}$ is existentially closed in $\mathfrak{B}$. Then, by Theorem 3.3.3(2), $\Sigma^*$ is model complete. It therefore remains to prove (ii); i.e. if $\mathfrak{A}$ is a model of $T$, then we need a $\mathfrak{B} \in E(T)$ with $\mathfrak{A} \subseteq \mathfrak{B}$. This follows from the inductiveness of $T$, using the next theorem. □

From Theorem 3.3.8(1), the uniqueness of a possible model companion $T^*$ of $T$ follows immediately. Namely, if $T'$ is another theory that is a model companion of $T$, then (1) yields

$$\mathrm{Mod}(T^*) = \mathrm{Mod}(T').$$

From this follows, using the Gödel Completeness Theorem 1.5.2,

$$T^* = \mathrm{Ded}(T^*) = \mathrm{Ded}(T') = T'.$$

**Theorem 3.3.9.** *Let $\mathscr{M}$ be an inductive class of L-structures. Then to every $\mathfrak{A} \in \mathscr{M}$ there is an $\mathfrak{A}^* \in E(\mathscr{M})$ with $\mathfrak{A} \subseteq \mathfrak{A}^*$.*

*Proof*: The construction of $\mathfrak{A}^*$ is similar to the construction in the proof of Theorem 2.5.2. We construct an $\omega$-chain $\big(\mathfrak{A}^{(n)}\big)_{n\in\mathbb{N}}$ in $\mathscr{M}$ with $\mathfrak{A}^{(0)} = \mathfrak{A}$.

If $\mathfrak{A}^{(n)} \in \mathscr{M}$ has already been defined, then let $\big(\varphi_\nu^{(n)}\big)_{\nu<\alpha_n}$ be an ordinal enumeration of all existential $L\big(\mathfrak{A}^{(n)}\big)$-sentences. We then set

$$\mathfrak{A}^{(n+1)} = \bigcup_{\nu<\alpha_n} \mathfrak{A}_\nu^{(n)},$$

where the $\alpha_n$-chain $\big(\mathfrak{A}_\nu^{(n)}\big)_{\nu<\alpha_n}$ is defined as follows:

$$\begin{aligned}
\mathfrak{A}_0^{(n)} &= \mathfrak{A}^{(n)}.\\
\mathfrak{A}_{\nu+1}^{(n)} &= \begin{cases} \text{any } \mathfrak{B} \in \mathscr{M} \text{ with } \mathfrak{A}_\nu^{(n)} \subseteq \mathfrak{B} \models \varphi_\nu^{(n)}, & \text{if this exists;}\\ \mathfrak{A}_\nu^{(n)} & \text{otherwise.}\end{cases}\\
\mathfrak{A}_\lambda^{(n)} &= \bigcup_{\nu<\lambda} \mathfrak{A}_\nu^{(n)} \text{ for limit ordinals } \lambda < \alpha_n.
\end{aligned}$$

By the inductiveness of $\mathscr{M}$, $\mathfrak{A}^{(n+1)}$ lies again in $\mathscr{M}$. Finally, we set

$$\mathfrak{A}^* = \bigcup_{n\in\mathbb{N}} \mathfrak{A}^{(n)}.$$

Now if $\varphi$ is an existential $L(\mathfrak{A}^*)$-sentence, then only finitely many constant symbols $\underline{a}$ with $a \in |\mathfrak{A}^*|$ occur in $\varphi$. Then $\varphi$ is already an $L(\mathfrak{A}^{(n)})$-sentence for some $n$, and therefore has an index $\nu < \alpha_n$. Let $\varphi$ be, say, $\varphi_\nu^{(n)}$. If $\varphi$ holds in an extension $\mathfrak{B} \in \mathscr{M}$ of $\mathfrak{A}^*$, then it clearly also holds in an extension of $\mathfrak{A}_\nu^{(n)}$ (namely, in $\mathfrak{B}$). But then $\varphi$ also holds in $\mathfrak{A}_{\nu+1}^{(n)}$, by definition. Because of the existential character of $\varphi$, we thus obtain, finally, the truth of $\varphi$ in $\mathfrak{A}^*$. Therefore $\mathfrak{A}^*$ is existentially closed in $\mathscr{M}$. □

## 3.4 Quantifier Elimination

In the previous section we concerned ourselves with theories in which every formula is equivalent to an $\forall$-formula. Now we turn our attention to the stronger condition that every formula is even equivalent to a quantifier-free formula. In order for this to be possible even for sentences of our language $L$, we always assume in this connection the existence of a(n individual) constant symbol, i.e. $K \neq \emptyset$. (Otherwise we must introduce a sentence-constant symbol $T$ that holds in every $L$-structure.) Then we say that $\Sigma \subseteq \mathrm{Sent}(L)$ admits *quantifier elimination* (briefly, "$\Sigma$ admits QE"), if to each $L$-formula $\varphi$ there is a quantifier-free $L$-formula $\delta$ with

$$\Sigma \vdash \forall(\varphi \leftrightarrow \delta) \quad \text{and} \quad \mathrm{Fr}(\delta) \subseteq \mathrm{Fr}(\varphi). \tag{3.4.0.1}$$

The proof of the quantifier elimination property usually proceeds with the help of the following criterion. Here we call a formula $\varphi$ a *simple existence formula* if it has the form $\exists x\, \delta$ with $\delta$ quantifier-free.

**Theorem 3.4.1.** *For $\Sigma \subseteq \mathrm{Sent}(L)$, the following are equivalent:*

(1) *$\Sigma$ admits quantifier elimination.*

(2) *$\Sigma \cup D(\mathfrak{A})$ is complete for every substructure $\mathfrak{A}$ of a model of $\Sigma$.*

(3) *For every two models $\mathfrak{B}_1$ and $\mathfrak{B}_2$ of $\Sigma$ with a common, finitely generated substructure $\mathfrak{A}$, and for every simple existence sentence $\varphi$ of the language $L(\mathfrak{A})$:*

$$(\mathfrak{B}_1, |\mathfrak{A}|) \overset{\varphi}{\rightsquigarrow} (\mathfrak{B}_2, |\mathfrak{A}|).$$

*Proof*: (1) $\Rightarrow$ (2): Let $\mathfrak{B}'_1$ and $\mathfrak{B}'_2$ be models of $\Sigma \cup D(\mathfrak{A})$. By Lemma 2.4.2, $\mathfrak{B}'_1$ and $\mathfrak{B}'_2$ have substructures isomorphic to $\mathfrak{A}$. We identify these with $\mathfrak{A}$. If we denote the $L$-restriction of $\mathfrak{B}'_\nu$ by $\mathfrak{B}_\nu$, for $\nu = 1,2$, then, writing $A = |\mathfrak{A}|$, we obviously have

$$\mathfrak{B}'_\nu = (\mathfrak{B}_\nu, A).$$

Let $\varphi'$ be an $L(\mathfrak{A})$-sentence. If we think of $\varphi'$ as arising from a suitable $L$-formula $\varphi$ by the substitution of constant symbols $\underline{a}$ for its free variables, and if $\delta$ is a quantifier-free formula satisfying (3.4.0.1), then we see that there is a quantifier-free $L(\mathfrak{A})$-sentence $\delta'$ with

$$\Sigma \vdash (\varphi' \leftrightarrow \delta').$$

Now, if $\varphi'$ holds in $(\mathfrak{B}_1, A)$, then so does $\delta'$. This has as consequences, first, $(\mathfrak{A}, A) \models \delta'$, and then $(\mathfrak{B}_2, A) \models \delta'$, since $\delta'$ is quantifier-free. This leads back to $(\mathfrak{B}_2, A) \models \varphi'$. Altogether, we therefore have

$$(\mathfrak{B}_1, A) \equiv (\mathfrak{B}_2, A).$$

Then by Lemma 3.2.1, $\Sigma \cup D(\mathfrak{A})$ is complete.

(2) $\Rightarrow$ (3) is trivial, since (3) is a special case of (2) (using Lemma 3.2.1).

(3) $\Rightarrow$ (1): Let $\varphi$ be an $L$-formula. Later we shall prove, by induction on the recursive construction of $\varphi$, the existence of a quantifier-free $L$-formula $\delta$ satisfying (3.4.0.1). But first we shall treat the special case in which $\varphi$ is of the form $\exists x \gamma$ with $\gamma$ quantifier-free.

Suppose $\mathrm{Fr}(\varphi) \subseteq \{v_0, \ldots, v_n\}$. We extend the language $L$ with new constants $c_0, \ldots, c_n$ (where $\{0, \ldots, n\} \cap K = \emptyset$). Let

$$\varphi' = \varphi(v_0/c_0, \ldots, v_n/c_n).$$

Call the resulting language $L'$. Let $\Gamma$ be the set of all quantifier-free sentences of $L'$. Further, let $\mathfrak{B}'_1$ and $\mathfrak{B}'_2$ be models of $\Sigma$, in the language $L'$, with $\mathfrak{B}'_1 \overset{\Gamma}{\rightsquigarrow} \mathfrak{B}'_2$. We wish to show $\mathfrak{B}'_1 \overset{\varphi}{\rightsquigarrow} \mathfrak{B}'_2$, in order to be able to apply Lemma 3.1.6.

Let $\mathfrak{B}_\nu$ be the restriction of $\mathfrak{B}'_\nu$ to $L$, for $\nu = 1,2$. Then the structure $\mathfrak{B}'_\nu$ has the form

$$\mathfrak{B}'_\nu = (\mathfrak{B}_\nu, (a_0^{(\nu)}, \ldots, a_n^{(\nu)})),$$

where $a_\mu^{(\nu)}$ is the interpretation of $c_\mu$ in $|\mathfrak{B}_\nu|$, for $0 \leq \mu \leq n$. Let $\mathrm{CT}'$ be the set of constant terms of $L'$. One can think of each term $t'$ in $\mathrm{CT}'$ as arising from an $L$-term $t$ by the substitution of $c_0, \ldots, c_n$ for the variables $v_0, \ldots, v_n$; i.e.

$$t' = t(v_0/c_0, \ldots, v_n/c_n).$$

The interpretation $t'^{\mathfrak{B}'_\nu}$ of $t'$ in $\mathfrak{B}'_\nu$ is, then, an element of $|\mathfrak{B}_\nu|$. We have

$$t'^{\mathfrak{B}'_\nu} = t^{\mathfrak{B}_\nu}\left[a_1^{(\nu)},\ldots,a_1^{(\nu)}\right].$$

Now we set

$$A_\nu := \left\{ t'^{\mathfrak{B}'_\nu} \,\middle|\, t' \in \mathrm{CT}' \right\}.$$

As we see immediately, $A_\nu$ is closed under all functions $f_j^{\mathfrak{B}_\nu}$ $(j \in J)$ of the structure $\mathfrak{B}_\nu$, and contains all constant interpretations $c_k^{\mathfrak{B}_\nu}$ $(k \in K)$. Therefore $A_\nu$ defines a substructure $\mathfrak{A}_\nu$ of $\mathfrak{B}_\nu$, by (2.3.0.1). Obviously, $\mathfrak{A}_\nu$ is finitely generated – namely, by $a_0^{(\nu)},\ldots,a_n^{(\nu)}$.

We define an isomorphism $\tau$ from $\mathfrak{A}_1$ to $\mathfrak{A}_2$ by

$$\tau\left(t'^{\mathfrak{B}'_1}\right) := t'^{\mathfrak{B}'_2}.$$

The well-definedness and the injectivity of the mapping $\tau$ are contained in the following equivalence ("$\Rightarrow$" is the well-definedness, and "$\Leftarrow$" is the injectivity). Let $t_1'$ and $t_2'$ be in $\mathrm{CT}'$. Then we have:

$$\begin{aligned} t_1'^{\mathfrak{B}'_1} = t_2'^{\mathfrak{B}'_1} \quad &\text{iff} \quad \mathfrak{B}'_1 \models t_1' \doteq t_2' \\ &\text{iff} \quad \mathfrak{B}'_2 \models t_1' \doteq t_2' \\ &\text{iff} \quad t_1'^{\mathfrak{B}'_2} = t_2'^{\mathfrak{B}'_2}. \end{aligned}$$

Here we have applied $\mathfrak{B}'_1 \overset{\Gamma}{\rightsquigarrow} \mathfrak{B}'_2$ to the quantifier-free $L'$-sentences $t_1' \doteq t_2'$ and $\neg\, t_1' \doteq t_2'$. The surjectivity of $\tau$ is trivial. It remains to prove the conditions ($I_2$), ($I_3$), and ($I_4$) of Section 2.2. Here, ($I_3$) and ($I_4$) are trivially fulfilled. ($I_2$) remains. So let $R_i$ be a relation symbol with $i \in I$. Then for $t_1',\ldots,t_{\lambda(i)}' \in \mathrm{CT}'$, we have:

$$\begin{aligned} R_i^{\mathfrak{B}'_1}\left(t_1'^{\mathfrak{B}'_1},\ldots,t_{\lambda(i)}'^{\mathfrak{B}'_1}\right) \quad &\text{iff} \quad \mathfrak{B}'_1 \models R_i\left(t_1',\ldots,t_{\lambda(i)}'\right) \\ &\text{iff} \quad \mathfrak{B}'_2 \models R_i\left(t_1',\ldots,t_{\lambda(i)}'\right) \\ &\text{iff} \quad R_i^{\mathfrak{B}'_2}\left(t_1'^{\mathfrak{B}'_2},\ldots,t_{\lambda(i)}'^{\mathfrak{B}'_2}\right). \end{aligned}$$

Here we have applied $\mathfrak{B}'_1 \overset{\Gamma}{\rightsquigarrow} \mathfrak{B}'_2$ to the quantifier-free $L'$-sentences $R_i\left(t_1',\ldots,t_{\lambda(i)}'\right)$ and $\neg R_i\left(t_1',\ldots,t_{\lambda(i)}'\right)$.

Thus we have, finally, found isomorphic, finitely generated substructures $\mathfrak{A}_1$ and $\mathfrak{A}_2$ of $\mathfrak{B}_1$ and $\mathfrak{B}_2$, respectively. If we identify these, then the sentence $\varphi'$ can be transferred from $\mathfrak{B}'_1$ to $\mathfrak{B}'_2$; i.e. $\mathfrak{B}'_1 \overset{\varphi'}{\rightsquigarrow} \mathfrak{B}'_2$. Using Lemma 3.1.6, we obtain a sentence $\delta' \in \Gamma$ with $\Sigma \vdash (\varphi' \leftrightarrow \delta')$. Since the interpretation of the new constant symbols $c_0,\ldots,c_n$ in the model $\mathfrak{B}$ of $\Sigma$ was completely arbitrary, we immediately obtain from this that

$$\Sigma \vdash \forall(\varphi \leftrightarrow \delta),$$

where $\delta = \delta'(c_0/v_0,\ldots,c_n/v_n)$ is a quantifier-free $L$-formula.

Thus we have taken care of the special case $\varphi = \exists x \gamma$ with $\gamma$ quantifier-free. Now we come to the induction on the recursive construction of $\varphi$.

If $\varphi$ is atomic, we take $\delta = \varphi$.

If $\varphi = \neg\varphi_1$ and if (by the inductive hypothesis) $\delta_1$ is quantifier-free with

$$\Sigma \vdash (\varphi_1 \leftrightarrow \delta_1) \qquad \text{and} \quad \mathrm{Fr}(\delta_1) \subseteq \mathrm{Fr}(\varphi_1), \tag{3.4.1.1}$$

then we take $\delta = \neg\delta_1$ and obtain what we need.

If $\varphi = (\varphi_1 \wedge \varphi_2)$ and if, by the inductive hypothesis, $\delta_1$ and $\delta_2$ are quantifier-free with

$$\Sigma \vdash (\varphi_\nu \leftrightarrow \delta_\nu) \qquad \text{and} \quad \mathrm{Fr}(\delta_\nu) \subseteq \mathrm{Fr}(\varphi_\nu)$$

for $\nu = 1,2$, then we take $\delta = (\delta_1 \wedge \delta_2)$ and again obtain what we need.

Finally, if $\varphi = \forall x \varphi_1$ and if, by the inductive hypothesis, $\delta_1$ is quantifier-free satisfying (3.4.1.1), then it follows, first, that

$$\Sigma \vdash (\forall x \varphi_1 \leftrightarrow \forall x \delta_1) \text{ and } \mathrm{Fr}(\forall x \delta_1) \subseteq \mathrm{Fr}(\forall x \varphi_1). \tag{3.4.1.2}$$

By the special case handled above, we obtain a quantifier-free $\delta$ with

$$\Sigma \vdash (\forall x \delta_1 \leftrightarrow \delta) \qquad \text{and} \quad \mathrm{Fr}(\delta) \subseteq \mathrm{Fr}(\forall x \delta_1). \tag{3.4.1.3}$$

But (3.4.1.2) and (3.4.1.3) combine to give

$$\Sigma \vdash (\forall x \varphi_1 \leftrightarrow \delta) \qquad \text{and} \quad \mathrm{Fr}(\delta) \subseteq \mathrm{Fr}(\forall x \varphi_1). \qquad \square$$

Because of property (2), a theory that admits quantifier elimination is also called *substructure complete* (by analogy with the definition of "model complete").

An $L$-structure $\mathfrak{P}$ that can be embedded in every model of an axiom system $\Sigma$ is called a *prime substructure* for $\Sigma$. From Theorem 3.4.1(2) we therefore obtain the following

**Corollary 3.4.2.** *If $\Sigma \subseteq \mathrm{Sent}(L)$ admits quantifier elimination, and if $\Sigma$ has a prime substructure, then $\Sigma$ is complete.* $\square$

From the definition of quantifier elimination (and with the help of Robinson's Test 3.3.3), or from Theorem 3.4.1(2), we see immediately that every theory admitting quantifier elimination is, naturally, model complete. The converse, as we shall see later, does not hold in general. The following theorem identifies the obstruction. Here we shall say that $\Sigma \subseteq \mathrm{Sent}(L)$ has the *amalgamation property* if, for every two models $\mathfrak{B}_1, \mathfrak{B}_2$ of $\Sigma$, and for every common substructure $\mathfrak{A}$ of $\mathfrak{B}_1$ and $\mathfrak{B}_2$, there are embeddings $\tau_\nu : \mathfrak{B}_\nu \to \mathfrak{B}$ $(\nu = 1,2)$ into a model $\mathfrak{B}$ of $\Sigma$ such that $\tau_1(a) = \tau_2(a)$ for all $a \in |\mathfrak{A}|$.

**Theorem 3.4.3.** $\Sigma \subseteq \mathrm{Sent}(L)$ *admits quantifier elimination if and only if* $\Sigma$ *is model complete and has the amalgamation property.*

*Proof*: First suppose $\Sigma$ is model complete and has the amalgamation property. We argue using Theorem 3.4.1(3). Let $\mathfrak{B}_1$ and $\mathfrak{B}_2$ be models of $\Sigma$ with the common substructure $\mathfrak{A}$. By hypothesis, there exist embeddings $\tau_\nu : \mathfrak{B}_\nu \to \mathfrak{B}$ $(\nu = 1,2)$ into a model $\mathfrak{B}$ of $\Sigma$ such that $\tau_1|_A = \tau_2|_A$, where $A = |\mathfrak{A}|$.

Now let $\varphi$ be a simple existential $L(\mathfrak{A})$-sentence with $(\mathfrak{B}_1, A) \models \varphi$. Because of the existential character of $\varphi$, we immediately deduce $(\mathfrak{B}, \tau_1|_A) \models \varphi$, which, in turn, leads to $(\mathfrak{B}, \tau_2|_A) \models \varphi$ using the hypothesis that $\tau_1|_A = \tau_2|_A$. From the model completeness of $\Sigma$, it follows that the embedding $\tau_2$ (and, naturally, also $\tau_1$) is elementary. We therefore obtain, finally, $(\mathfrak{B}_2, A) \models \varphi$. We have thus proved criterion (3) of Theorem 3.4.1.

Now suppose $\Sigma$ admits quantifier elimination. It remains to prove the amalgamation property. So let $\mathfrak{B}_1$ and $\mathfrak{B}_2$ be models of $\Sigma$ with the common substructure $\mathfrak{A}$. Then, with the help of Theorem 2.5.2, we choose a $\kappa$-saturated, elementary extension $\mathfrak{B}$ of $\mathfrak{B}_2$, where we may take $\kappa > \mathrm{card}(|\mathfrak{B}_1|)$. Then by Theorem 2.5.4, the structure $(\mathfrak{B}_1, |\mathfrak{A}|)$ can be embedded in $(\mathfrak{B}, |\mathfrak{A}|)$, provided every $\exists$-sentence of the $L(\mathfrak{A})$-language that holds in $(\mathfrak{B}_1, |\mathfrak{A}|)$ also holds in $(\mathfrak{B}, |\mathfrak{A}|)$. But this is guaranteed by criterion (2) of Theorem 3.4.1. We therefore have an embedding $\tau_1 : \mathfrak{B}_1 \to \mathfrak{B}$ that is the identity on $|\mathfrak{A}|$. On the other hand, the identity mapping is naturally an embedding of $\mathfrak{B}_2$ into $\mathfrak{B}$; in particular, this embedding agrees with $\tau_1$ on $|\mathfrak{A}|$. As an elementary extension of $\mathfrak{B}_2$, $\mathfrak{B}$ is, finally, also a model of $\Sigma$. □

We observe, moreover, that in the last proof, when proving quantifier elimination under the hypothesis of the amalgamation property, we did not exploit the full strength of the model completeness hypothesis on $\Sigma$, but rather only the following – in general not equivalent – consequence of model completeness: if $\mathfrak{A}$ and $\mathfrak{B}$ are models of $\Sigma$ with $\mathfrak{A} \subseteq \mathfrak{B}$, then every *simple* existential $L(\mathfrak{A})$-sentence that holds in $(\mathfrak{B}, |\mathfrak{A}|)$ already holds in $(\mathfrak{A}, |\mathfrak{A}|)$. This observation can in some cases significantly simplify the proof of quantifier elimination in the presence of the amalgamation property.

Before bringing this section to a close, we would like to prove quantifier elimination for the theory of algebraically closed fields, without falling back on earlier results based on the theorems of Steinitz. By means of criterion (3) of Theorem 3.4.1, we shall show that the algebraic facts that we shall need to bring in are very simple. On the other hand, however, from quantifier elimination will immediately follow model completeness for this theory (Theorem 3.3.4), and, for fixed characteristic, using Corollary 3.4.2, also completeness (Theorem 3.2.3(3)). As for the hypothesis of Corollary 3.4.2, we need observe only that the prime subfield $\mathbb{Q}$, or $\mathbb{F}_p$ for a prime number $p$, is obviously a prime substructure.

**Theorem 3.4.4.** *The theory of algebraically closed fields admits quantifier elimination.*

*Proof*: We shall prove condition (3) of Theorem 3.4.1. For this, let $F_1$ and $F_2$ (in the usual notation for fields) be algebraically closed fields, and $\mathfrak{A}$ a common substructure. It is clear that $\mathfrak{A}$ must be a subring of $F_1$ and $F_2$. We may therefore assume that the quotient field $F$ of $\mathfrak{A}$ is likewise contained in $F_1$ and $F_2$. It is well known that the algebraic closure $\widetilde{F}$ of a field $F$ is, up to isomorphism, uniquely determined. We may therefore assume, further, that $\widetilde{F}$ is contained in $F_1$ and $F_2$; i.e. we have the situation

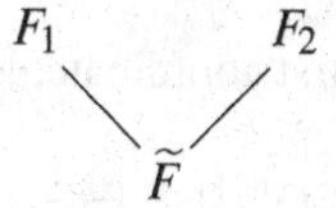

where all three fields are algebraically closed.

Let $\exists x\,\delta$ be an $L(A)$-sentence with $\delta$ quantifier-free and $(F_1,A) \models \exists x\,\delta$, where $A = |\mathfrak{A}|$. We must show $(F_2,A) \models \exists x\,\delta$. For simplicity, in the following we shall write only $F_\nu$ for $(F_\nu,A)$.

Without loss of generality we can assume that $\delta$ is given in disjunctive normal form, i.e.

$$\delta = (\delta_1 \vee \dots \vee \delta_m),$$

where $\delta_\nu$ is a conjunction of atomic formulae and negated atomic formulae. Since $\exists x\,\delta$ is logically equivalent to

$$(\exists x\,\delta_1 \vee \dots \vee \exists x\,\delta_m),$$

at least one disjunct $\exists x\,\delta_\nu$ must hold in $F_1$. If we can show $F_2 \models \exists x\,\delta_\nu$, then $F_2 \models \exists x\,\delta$ will naturally follow.

We may further assume that $\delta_\nu$ has the following form:

$$\left(\bigwedge_{i=1}^{r} p_i(x) \doteq 0 \wedge \bigwedge_{j=1}^{s} q_j(x) \neq 0\right)$$

(where the cases $r = 0$ or $s = 0$ are not excluded). Here, $p_i$ and $q_j$ are polynomials in $x$, whose coefficients are themselves polynomials in the constants $\underline{a_1}, \dots, \underline{a_n}$ occurring in $\delta$, with integer coefficients. More precisely, these "integer" coefficients are terms that may be taken to be of the form $\pm(1 + \dots + 1)$.

If the interpretations of the coefficients of the terms $p_i$ $(1 \le i \le r)$ yield at least one nonzero polynomial in $A[x]$, then any element $b \in F_1$ that satisfies $\delta_\nu$ in $F_1$ must lie already in $\widetilde{F}$. Thus $\widetilde{F} \models \exists x\,\delta_\nu$, whence, *a fortiori*, $F_2 \models \exists x\,\delta_\nu$. If, on the other hand, all the terms $p_i$ yield only the identically zero polynomial, or if $r = 0$, then it suffices to find an element $b \in \widetilde{F}$ that differs from all the zeros of polynomials in $A[x]$ determined by the terms $q_j$. One such $b$ exists, since none of the terms $q_j(x)$ can yield the identically zero polynomial in $x$ (so that there are only finitely many such zeros), and $\widetilde{F}$, being an algebraically closed field, has infinite cardinality. We therefore have $F_2 \models \exists x\,\delta_\nu$, and hence again $F_2 \models \exists x\,\delta$. □

As an application of quantifier elimination for the theory of algebraically closed fields $F$, one gets, e.g. that for $F$, the projections of constructible subsets of the affine space $F^n$ are again constructible. Here we call a subset $M$ of $F^n$ *constructible* if it is a Boolean combination of algebraic sets. In our language, this means that

$M$ is defined by a quantifier-free formula; more precisely, there is a quantifier-free formula $\delta$ of the language $L(F)$ with $\mathrm{Fr}(\delta) \subseteq \{x_1,\ldots,x_n\}$, such that

$$M = \{ (a_1,\ldots,a_n) \in F^n \mid F \models \delta(\underline{a_1},\ldots,\underline{a_n}) \}.$$

A projection of $M$, e.g. along the first coordinate, is then just the set

$$M_1 = \{ (a_2,\ldots,a_n) \in F^{n-1} \mid F \models \exists x_1\, \delta(x_1,\underline{a_2},\ldots,\underline{a_n}) \}.$$

Since $\exists x_1\, \delta$ is can be replaced by an equivalent quantifier-free formula $\delta_1$ with $\mathrm{Fr}(\delta_1) \subseteq \{x_2,\ldots,x_n\}$, we see that $M_1$ is again constructible.

## 3.5 Exercises for Chapter 3

**3.5.1.** Let $L = (+,\cdot,0,1)$, and let $\mathrm{Mod}_L$ be the class of all $L$-structures. We endow $\mathrm{Mod}_L$ with the topology generated by the basis consisting of sets of the form

$$\mathrm{Mod}_L(\sigma) := \{\mathfrak{A} \in \mathrm{Mod}_L \mid \mathfrak{A} \models \sigma\},$$

for $\sigma \in \mathrm{Sent}(L)$.

For each of the following subclasses of $\mathrm{Mod}_L$, determine whether it is open, closed, both or neither:

(a) the class of fields;
(b) the class of fields with fixed characteristic $p$ (for $p$ a prime number or 0);
(c) the class of fields with nonzero characteristic; and
(d) the class of factorial rings.

**3.5.2.** Let $L$ be a fixed language, $R$ a new $n$-ary relation symbol, and $L_R$ the language $L$ together with $R$. Assume that $\Sigma_R$ is an axiom system in the $L_R$-language such that for every $L$-isomorphism of two models $\mathfrak{A}_1$ and $\mathfrak{A}_2$ of $\Sigma_R$, $R^{\mathfrak{A}_1}$ maps to $R^{\mathfrak{A}_2}$. Then $R$ is $L$-definable with respect to $\Sigma_R$, i.e. there exists an $L$-formula $\varphi$ with $\mathrm{Fr}(\varphi) \subseteq \{x_1,\ldots,x_n\}$ such that

$$\Sigma_R \vdash \forall x_1,\ldots,x_n\, (R(x_1,\ldots,x_n) \leftrightarrow \varphi(x_1,\ldots,x_n)).$$

*Hint:* Apply Lemma 3.1.6, and recall Theorems 2.5.2 and 2.5.7 (using GCH), or Exercise 2.7.14 (not using GCH).

**3.5.3.** Consider $L, R$ and $\Sigma_R$ from Exercise 3.5.2. Let $R'$ be an $n$-ary relation symbol, new with respect to $L_R$. Denote by $\Sigma_{R'}$ the axiom system obtained from $\Sigma_R$ by replacing $R$ with $R'$. We call $R$ *implicitly definable* with respect to $\Sigma_R$ if

$$\Sigma_R \cup \Sigma_{R'} \vdash \forall x_1,\ldots,x_n\, (R(x_1,\ldots,x_n) \leftrightarrow R'(x_1,\ldots,x_n)).$$

From Exercise 3.5.2 deduce *Beth's Theorem*, which says that if $R$ is implicitly definable with respect to $\Sigma_R$, then it is already (explicitly) definable with respect to $\Sigma_R$.

**3.5.4.** Prove the following statements:

(a) The theory of dense linear orderings without extrema is not $\aleph_1$-categorical.

(b) The theory of algebraically closed fields of a fixed characteristic is not $\aleph_0$-categorical.

(c) The theory of divisible, ordered, Abelian groups is neither $\aleph_0$- nor $\aleph_1$-categorical.

**3.5.5.** Let $\Sigma \subseteq \text{Sent}(L)$. An axiom system $\Sigma^* \subseteq \text{Sent}(L)$ is called a *model companion* of $\Sigma$ if:

(i) every model of $\Sigma^*$ embeds in a model of $\Sigma$,
(ii) every model of $\Sigma$ embeds in a model of $\Sigma^*$,
(iii) $\Sigma^*$ is model complete.

Prove the following facts:

(a) A strict model companion of $\Sigma$ (defined on p. 118) is also a model companion.
(b) A model companion $\Sigma^*$ of $\Sigma$ is uniquely determined (if it exists).
(c) If $\text{Mod}(\Sigma)$ is inductive, then a model companion is already strict.

**3.5.6.** Show that the theory of "proper" integral domains (i.e. those that are not already fields) admits a model companion that is not strict.

**3.5.7.** Show that the theory $T$ of commutative rings $R = \langle R; +^R, -^R, \cdot^R; 0^R, 1^R\rangle$ with 1 (axiomatized by $K_1$–$K_6$,$K_8$,$K_9$) has no (strict) model companion.

*Hint*: Show that in every existentially closed model $R$ of $T$:

(a) $x \in R$ is nilpotent if and only if $R \models \forall y((xy)^2 \doteq xy \rightarrow xy \doteq 0)$, and

(b) there are nilpotent elements in $R$ of every order of nilpotency, i.e. to every $n \in \mathbb{N} \setminus \{0\}$ there is an $x \in R$ with $x^n \neq 0$ but $x^{n+1} = 0$.

From this deduce that the class $E(T)$ is not axiomatizable. Now apply Theorem 3.3.8(1).

**3.5.8.** Let $L = (+, 0)$. Show that the axioms for Abelian groups do not admit quantifier elimination.

# Chapter 4
# Model Theory of Several Algebraic Theories

In this chapter we investigate a series of interesting algebraic theories for the properties of completeness, model completeness and quantifier elimination. Not only do we treat the standard examples that have already been frequently treated in the extant literature, but we shall place special value on the theory of valued fields. Since valuation theory does not belong in the standard repertoire of an algebra course, we shall first discuss the necessary concepts and theorems in detail, in Section 4.3. Thereafter we develop special cases (Sections 4.4 and 4.5), and finally the model theory of Henselian valued fields. The goal of this presentation is, among other things, a treatment of a purely number theoretic problem – Artin's conjecture – in Theorem 4.6.5.

## 4.1 Ordered Abelian Groups

We begin the model theoretic investigation of algebraic theories with the theory of ordered Abelian groups. As an axiom system of this theory we shall use

$$\{\mathrm{G}_1, \mathrm{G}_2, \mathrm{G}_3, \mathrm{G}_4\} \cup \{\mathrm{O}_1, \mathrm{O}_2, \mathrm{O}_3, \mathrm{OG}\} \cup \{\exists \mathrm{x}\ \mathrm{x} \neq 0\}$$

in a language $L$ with the usual symbols $<$, $+$, $0$ (recall Section 1.6, and note that here we do not consider the trivial group as an ordered group). Then models have the form

$$\mathfrak{A} = \langle A; <^{\mathfrak{A}}; +^{\mathfrak{A}}; 0^{\mathfrak{A}} \rangle.$$

For this we also write, simply,

$$\mathfrak{A} = \langle A; <; +; 0 \rangle$$


DOI 10.1007/978-1-4471-2176-3_5, 

in what follows; i.e. we denote the interpretations in $\mathfrak{A}$ of the symbols $<$, $+$, and $0$ again by $<$, $+$, and $0$ themselves. We shall usually denote ordered Abelian groups by the letter $G$.

If $G$ is an ordered Abelian group, then $G$ may possess a smallest positive element; i.e. the following sentence may hold in $G$:

$$\text{DO:}\quad \exists x\,(0 < x \wedge \forall y\,(0 < y \to x \doteq y \vee x < y)).$$

In this case we called $G$ *discrete ordered* in Section 1.6(4). The integers $\mathbb{Z}$ with their usual ordering furnish an example of this. If, on the other hand, $G$ has no smallest positive element, then $G$ is *densely ordered* (i.e. $O_4$ of Section 1.6(1) holds). Indeed, if $a, b \in |G|$ with $a < b$, and if $0 < c < b - a$, then it naturally follows that

$$a < c + a < b.$$

Now, in the next two theorems, we would like to investigate two special classes of ordered Abelian groups. We begin with

**Theorem 4.1.1.** *The theory of divisible, ordered, Abelian groups admits quantifier elimination, and is complete and model complete.*

*Proof*: Model completeness follows immediately from quantifier elimination; and completeness follows from Corollary 3.4.2, since obviously $\mathbb{O} = \langle \{0\};\ \emptyset;\ +;\ 0 \rangle$ with $0 + 0 = 0$ is a prime substructure.

To prove quantifier elimination, we proceed just as we did in the proof of Theorem 3.4.4. Let $G_1$ and $G_2$ be divisible ordered groups, and $H$ a common substructure. $H$ is a semigroup with the cancellation property. (Observe that the inverse operation is not in our language.) However, $H$ equals $\mathbb{O}$ or generates isomorphic ordered subgroups in $G_1$ and $G_2$; even their divisible hulls in $G_1$ and $G_2$ are still isomorphic. We identify them, and denote them by $\widetilde{H}$. We therefore have the following situation:

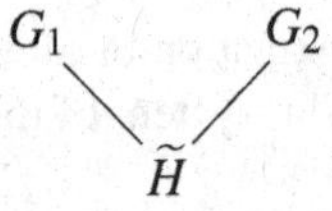


Here the case $H = \widetilde{H} = \mathbb{O}$ is not excluded.

As in the proof of Theorem 3.4.4, it suffices to consider the case in which $\delta_v$ is a conjunction of atomic formulae or negated atomic formulae of the language $L(H)$, and that at most $x$ is free in $\delta_v$. We assume $G_1 \models \exists x\,\delta_v$, and show $G_2 \models \exists x\,\delta_v$. For this, however, we can still make several simplifications in $\delta_v$: we replace the formulae

$$\begin{array}{lll} \neg(t_1 \doteq t_2) & \text{by} & (t_1 < t_2 \vee t_2 < t_1) \quad \text{and} \\ \neg(t_1 < t_2) & \text{by} & (t_2 < t_1 \vee t_1 \doteq t_1). \end{array} \tag{4.1.1.1}$$

For ordered groups, this is an equivalent replacement. If we then bring the result again into disjunctive normal form, and distribute the existential quantifier $\exists x$ over each disjunct, then we may assume $\delta_v$ to be of the form

$$\left(\bigwedge_{i=1}^{r} n_i x + \underline{a_i} \doteq n'_i x + \underline{a'_i} \wedge \bigwedge_{j=1}^{s} m_j x + \underline{b_j} < m'_j x + \underline{b'_j}\right).$$

Here we have already carried out possible group theoretic transformations, and introduced new constants (e.g. replacement of $\underline{a}+\underline{b}$ by $\underline{a+b}$). As usual, $nx$ denotes the $n$-fold sum $x+\cdots+x$ when $n>0$. The cases $r=0$ or $s=0$ is allowed. All constants $\underline{a}$ belong to elements $a \in |H|$.

Now, if some equation is nontrivial, i.e. $n_i \neq n'_i$, then any element $d$ in $G_1$ that satisfies $\delta_v$ lies already in $\widetilde{H}$. We therefore have $\widetilde{H} \models \exists x\,\delta_v$, and hence also $G_2 \models \exists x\,\delta_v$. If all equations are trivial, i.e. $n_i = n'_i$ for $1 \leq i \leq r$, or if $r=0$, then an element $d$ of $G_1$ obviously satisfies the formula $\delta_v$ if and only if

$$\begin{aligned} a_i &= a'_i && \text{for} \quad 1 \leq i \leq r \quad \text{and} \\ (m_j - m'_j)d &< b'_j - b_j && \text{for} \quad 1 \leq j \leq s. \end{aligned}$$

If $G_1$ contains such an element, then so does $\widetilde{H}$, clearly, in case $\widetilde{H} \neq \mathbb{O}$. And in this case, $\widetilde{H} \models \exists x\,\delta_v$ holds, and hence also $G_2 \models \exists x\,\delta_v$. On the other hand, if $\widetilde{H} = \mathbb{O}$, then $b_j = b'_j$ for $1 \leq j \leq s$. Then $\exists x\,\delta_v$ asserts merely (1) $a_i = a'_i$ and (in case $s>0$) (2) the existence of an element that is positive or negative, which naturally holds in $G_2$, too. □

**Corollary 4.1.2.** *The theory of divisible, ordered, Abelian groups is the model companion of the theory of ordered Abelian groups.*

*Proof*: Upon applying Theorem 4.1.1, it remains only to show that each ordered Abelian group $G$ can be embedded in a divisible, ordered, Abelian group. We shall show that $G$ can be embedded in its so-called *divisible hull*. This can be defined by analogy with the construction of $\mathbb{Q}$ from $\mathbb{Z}$, as follows. We consider pairs $(a,n)$, where $a \in |G|$ and $n \in \mathbb{N} \setminus \{0\}$ (here we think of quotients $\frac{a}{n}$). We declare two pairs $(a,n)$ and $(a',n')$ to be equivalent if $n'a = na'$. Now one may easily check that the equivalence classes of such pairs form a divisible, ordered, Abelian group $\widetilde{G}$, once we define addition in $\widetilde{G}$ by

$$(a,n)+(a',n') = (n'a+na', nn'),$$

and an ordering on $\widetilde{G}$ by

$$(a,n) < (a',n') \quad \text{iff} \quad n'a < na'.$$

$G$ embeds in $\widetilde{G}$ via $a \mapsto (a,1)$. Thus $\widetilde{G}$ is a divisible extension of $G$ with the property: to each $b \in |\widetilde{G}|$ there is an $a \in |G|$ and an $n \in \mathbb{N} \setminus \{0\}$ with $nb = a$. Moreover, $\widetilde{G}$ is, up to isomorphism, the *unique* divisible, ordered, Abelian extension group of $G$ with this property. □

In the next theorem we shall show, among other things, that the theory of $\mathbb{Z}$-groups is model complete. When axiomatizing this theory in Section 1.6, we added a constant symbol 1 to the language of ordered groups. The restriction of $\mathbb{Z}$-groups to the language of ordered groups can also be axiomatized: we need only describe 1 implicitly. The axioms are, then, the axioms given above for discrete, ordered, Abelian groups, together with

$$\mathrm{D}_n^- : \quad \forall u (0 < u \wedge \forall y (0 < y \rightarrow u \doteq y \vee u < y) \rightarrow \mathrm{D}_n(1/u)).^1$$

In each model of this axiom system $\Sigma$, there is a smallest positive element – the interpretation of 1 in the extended language. Obviously, $\mathbb{Z}$ and $2\mathbb{Z}$ are both models of $\Sigma$, and $2\mathbb{Z}$ is a substructure of $\mathbb{Z}$ (in the language of ordered groups). However, this is no longer true in the language extended by 1, since that symbol would be interpreted in $2\mathbb{Z}$ as 2, in order to obtain a $\mathbb{Z}$-group. We therefore have here a case in which the substructure relation changes upon extending the language.

In the following theorem, we shall show that the theory of $\mathbb{Z}$-groups is complete and model complete. From this, the completeness of the theory just described (without 1) will also follow. We thereby obtain an example of a complete theory that is not model complete. (Indeed, in the language without 1, we have $2\mathbb{Z} \subseteq \mathbb{Z}$, but not $2\mathbb{Z} \preceq \mathbb{Z}$.)

**Theorem 4.1.3.** *The theory of $\mathbb{Z}$-groups is model complete and complete.*

*Proof*: Completeness follows from model completeness by Corollary 3.3.2, since $\mathbb{Z}$ is obviously a prime model. We shall obtain model completeness via a detour through quantifier elimination in a suitable extension language.

Let $L$ be the language in which we axiomatized the theory of $\mathbb{Z}$-groups in Section 1.6. We extend this language by the addition of a unary relation $R_n$ for every $n \geq 2$ (without loss of generality, let $I \cap \mathbb{N} = \emptyset$). Then we extend the axiom system $\Sigma$ for $\mathbb{Z}$-groups to $\Sigma'$, by adding the axioms

$$\mathrm{A}_n : \quad \forall x (R_n(x) \leftrightarrow \exists y\, x \doteq ny).$$

Thus, the relation $R_n$ expresses divisibility by $n$. The axioms $\mathrm{D}_n$ can thus be reformulated as

$$\mathrm{D}'_n : \quad \forall x \bigvee_{\nu=1}^{n} R_n(x + \nu 1).$$

The restrictions to $L$ of models of $\Sigma'$ are exactly the $\mathbb{Z}$-groups. Observe that the substructure relation between models $\mathfrak{A}$ and $\mathfrak{B}$ of $\Sigma$ continues to hold when we extend these to models $\mathfrak{A}'$ and $\mathfrak{B}'$ of $\Sigma'$ by interpreting the relations $R_n$ according to the axioms $\mathrm{A}_n$. Namely, if for an element $a \in |\mathfrak{A}|$, the statement

$$a \equiv \nu \bmod n$$

[1] Recall, the notation $\mathrm{D}_n(1/u)$ was defined in (1.6.1.1) and (1.2.0.9).

holds in $\mathfrak{B}$ (i.e. if $a - v1$ is divisible by $n$ in $\mathfrak{B}$), then this must also hold in $\mathfrak{A}$, since $a$ is congruent to a $v'$ modulo $n$ also in $\mathfrak{A}$. This $v'$ must inevitably be equal to $v$.

We obtain the language $L'$ by further adding a unary function symbol, which we interpret as the additive inverse operation. Because of this, $\Sigma'$ contains also the axiom

$$\mathrm{K}_3: \quad \forall x\, x + (-x) = 0.$$

This extension does not alter the substructure relation between models, obviously.

Now we show that $\Sigma'$ admits quantifier elimination. $\Sigma'$ will thus be model complete, which, because of the invariance of the substructure relation, will imply the model completeness of $\Sigma$.

We shall prove the quantifier elimination of $\Sigma'$ directly in this case, i.e. we will not use Theorem 3.4.1, but rather successively transform a formula $\varphi$ into formulae equivalent to it modulo $\Sigma'$, until we obtain a quantifier-free formula $\delta$; at every stage, the set of free variables will contain at most those of $\varphi$. Here we say that $\alpha$ is "equivalent to $\beta$ modulo $\Sigma'$" if $\Sigma' \vdash (\alpha \leftrightarrow \beta)$, i.e. if in all models $\mathfrak{A}$ of $\Sigma'$ and for every evaluation $h$, we have:

$$\mathfrak{A} \models \alpha\,[h] \quad \text{iff} \quad \mathfrak{A} \models \beta\,[h].$$

Obviously it will suffice if, for every quantifier-free formula $\delta$ and every variable $x$, we can find a quantifier-free formula $\delta_1$ such that $\exists x\, \delta$ is equivalent to $\delta_1$ modulo $\Sigma'$. Then one continues as in the proof of Theorem 3.4.1, (3)$\Rightarrow$(1), by induction on the recursive construction of formulae.

It suffices, furthermore, to consider only unnegated occurrences of the relations $R_n(t)$ in $\delta$. Namely, we can, by $\mathrm{D}'_n$, replace $\neg R_n(t)$ by

$$R_n(t+1) \vee \cdots \vee R_n(t + (n-1)1),$$

which is equivalent to it modulo $\Sigma'$. We replace negations of $t_1 \doteq t_2$ and $t_1 < t_2$ as in (4.1.1.1). Then, if we bring $\delta$ into disjunctive normal form, and distribute the existential quantifier $\exists x$ over the individual disjuncts, we see that the only case yet to be treated is that in which $\delta$ is of the form

$$\left( \bigwedge_{i=1}^{r} m_i x \doteq t_i \wedge \bigwedge_{i=1}^{r'} m'_i x < t'_i \wedge \bigwedge_{i=1}^{r''} m''_i x \equiv t''_i \bmod n_i \right),$$

where $m_i, m'_i, m''_i \in \mathbb{Z}$ and

$$m''_i x \equiv t''_i \bmod n_i \quad \text{stands for} \quad R_{n_i}(m''_i x + (-t''_i)).$$

The cases $r = 0$, $r' = 0$, or $r'' = 0$ are not excluded.

One sees easily that we can arrange, by multiplication by a suitable integer ($\neq$ 0), that all coefficients $m_i, m'_i, m''_i$ are equal to a single integer $m$. (In the case of a congruence, one must, naturally, multiply also the modulus $n_i$ by the same integer!)

We can, therefore, assume that $\delta$ is of the form

$$\left(\bigwedge_{i=1}^{r} mx \doteq t_i \wedge \bigwedge_{i=1}^{r'} mx < t'_i \wedge \bigwedge_{i=1}^{r''} mx \equiv t''_i \bmod n_i\right).$$

If we replace $mx$ by $y$, and add the condition $y \equiv 0 \bmod m$, then we again obtain an equivalent formula modulo $\Sigma'$. After this step, we can, finally, assume that $\delta$ has the form

$$\left(\bigwedge_{i=1}^{r} x \doteq t_i \wedge \bigwedge_{i=1}^{r^+} x < t_i^+ \wedge \bigwedge_{i=1}^{r^-} t_i^- < x \wedge \bigwedge_{i=1}^{r''} x \equiv t''_i \bmod n_i\right),$$

where we again write $x$ for $y$. If at least one equation $x = t_i$ occurs (i.e. if $r > 0$), then $\exists x\, \delta$ is obviously equivalent, modulo $\Sigma'$, to

$$\left(\bigwedge_{i=2}^{r} t_1 \doteq t_i \wedge \bigwedge_{i=1}^{r^+} t_1 < t_i^+ \wedge \bigwedge_{i=1}^{r^-} t_i^- < t_1 \wedge \bigwedge_{i=1}^{r''} t_1 \equiv t''_i \bmod n_i\right).$$

In this case, we have eliminated the quantifier $\exists x$. If, on the other hand, no equation occurs (i.e. if $r = 0$), but at least one inequality $x < t_i^+$ occurs, then, letting $n$ be the least (positive) common multiple of $n_1, \ldots, n_{r''}$,[2] we observe that in every model $\mathfrak{A}$ of the system

$$x \equiv t''_i \bmod n_i \quad (1 \leq i \leq r''),$$

if $x$ is in $\mathfrak{A}$, then so is $x + ln$ for all $l \in \mathbb{Z}$. Setting

$$t^+ = \min_{1 \leq i \leq r^+} t_i^+,$$

we can obviously select $x$ from the set

$$\{t^+ - 1, t^+ - 2, \ldots, t^+ - n\}.$$

Therefore $\exists x\, \delta$ can, modulo $\Sigma'$, be expressed without quantifiers, as follows:

$$\bigvee_{k=1}^{r^+} \bigvee_{j=1}^{n} \left(\bigwedge_{i=1}^{r^+} t_k^+ \leq t_i^+ \wedge \bigwedge_{i=1}^{r^-} t_i^- < t_k^+ - j \wedge \bigwedge_{i=1}^{r''} t_k^+ - j \equiv t''_i \bmod n_i\right).$$

Next, in the case where $r = 0$ and $r^+ = 0$, but $r^- \geq 1$, we consider analogously (in a model $\mathfrak{A}$) the set

$$\{t^- + 1, \ldots, t^- + n\},$$

where

$$t^- = \max_{1 \leq i \leq r^-} t_i^-.$$

In this case, we can describe $\exists x\, \delta$ without quantifiers, by

[2] If $r'' = 0$, then we take $n = 1$.

$$\bigvee_{k=1}^{r^-}\bigvee_{j=1}^{n}\left(\bigwedge_{i=1}^{r^-} t_i^- \leq t_k^- \wedge \bigwedge_{i=1}^{r''} t_k^- + j \equiv t_i'' \bmod n_i\right).$$

Finally, if $r = r^+ = r^- = 0$, then, modulo $\Sigma'$, $\exists x\, \delta$ is equivalent to

$$\bigvee_{j=1}^{n}\left(\bigwedge_{i=1}^{r''} j \equiv t_i'' \bmod n_i\right). \qquad \square$$

**Corollary 4.1.4.** *The theory of* $\mathbb{Z}$*-groups (in the language with* 1*) is the model companion of the theory of discrete, ordered, Abelian groups.*

*Proof*: Upon application of Theorem 4.1.3, it remains only to show that every discrete, ordered, Abelian group $G$ can be extended to a $\mathbb{Z}$-group. For this, we work in the ordered divisible hull $\widetilde{G}$ of $G$ (compare the proof of Corollary 4.1.2).

Let $H$ be a maximal extension of $G$ in $\widetilde{G}$ such that $1^G$ is still the minimal positive element of $H$. The existence of $H$ is guaranteed by Zorn's lemma. Further, let $p$ be a prime number. If $a \in H$ is not divisible (in $H$) by $p$, then the extension $H + \mathbb{Z}\frac{a}{p}$ of $H$ no longer has $1^G$ as minimal positive element, by maximality. Thus, there is a $b \in H$ and an $m \in \mathbb{Z}$ not divisible by $p$ such that

$$0 < b + m\frac{a}{p} < 1.$$

(From now on we write simply $n$ for $n1^G$.) From this follows

$$0 < pb + ma < p. \tag{4.1.4.1}$$

Since $m$ and $p$ are relatively prime, there are $s, t \in \mathbb{Z}$ with

$$sp + tm = 1. \tag{4.1.4.2}$$

In the case where $0 < t$, (4.1.4.1) and (4.1.4.2) yield

$$0 < tpb + (1 - sp)a < pt,$$

or, upon rearranging,

$$0 < p(tb - sa) + a < pt.$$

By the discreteness of $H$, it follows that

$$p(tb - sa) + a = r,$$

for some $r \in \mathbb{Z}$. This follows also in the case where $t < 0$. Therefore,

$$a - r = pa_1, \tag{4.1.4.3}$$

for some $a_1 \in H$. Note that we have derived (4.1.4.3) from the hypothesis that $p$ does not divide $a$ in $H$; but (4.1.4.3) continues to hold, trivially, also when this hypothesis is not satisfied (namely, take $r = 0$). Therefore, for the above $a_1$ and for any prime number $q$, this process can be repeated – say,

$$a_1 - r_1 = qa_2, \tag{4.1.4.4}$$

for some $a_2 \in H$ and $r_1 \in \mathbb{Z}$. Combining (4.1.4.3) and (4.1.4.4), we get

$$a - r' = pqa_2a_1,$$

for some $r' \in \mathbb{Z}$. It is now clear that for every $n \geq 2$, we can find an $r \in \mathbb{Z}$ with

$$a - r \in nH,$$

where we may, in addition, assume that $0 \leq r < n$ still holds. We have therefore shown that $H$ is a $\mathbb{Z}$-group. □

## 4.2 Ordered Fields

Now we wish to prove, finally, the completeness of the theory of real closed fields – the last of the theories whose completeness we had, in Section 1.6, asserted. We begin with a short (but not complete) introduction to the algebraic theory of ordered fields. For literature on this, we refer to [Prestel–Delzell, 2001].

Let $F$ be a field. An ordering on $F$ can be given either by a binary relation $<$ on $F$ satisfying the axioms $O_1$, $O_2$, $O_3$, $OK_1$ and $OK_2$ (from Section 1.6), or by a *positive cone*: this is a subset $P$ of $F$ with the following properties:

$$P + P \subseteq P, \quad P \cdot P \subseteq P, \quad P \cap -P = \{0\}, \quad P \cup -P = F.$$

The connection between these two presentations is, then, the following: if $<$ is an ordering on $F$, then

$$P_< := \{a \in F \mid 0 \leq a\}$$

is a positive cone. (More precisely, $P_<$ consists of the *nonnegative* elements of $F$!) Conversely, if $P$ is a positive cone of $F$, then

$$a <_P b \quad \text{iff} \quad b - a \in P \setminus \{0\}$$

defines an ordering on $F$.

In the following we shall always write $(F, <)$ for an ordered field. Here $F$ is a field, and $<$ is an ordering on $F$. The class of all ordered fields is obviously closed under union of chains; i.e. it is inductive. Consequently, given an ordered field $(F, <)$, if we consider all extensions $(F', <')$ of it such that $F'$ is algebraic over $F$, then we obtain, using Zorn's lemma, the existence of a maximal ordered, algebraic extension $(F^*, <^*)$ of $(F, <)$. These extensions $(F^*, <^*)$ are real closed, by the following theorem:

**Theorem 4.2.1.** *For an ordered field $(F, <)$, the following are equivalent:*

(1) *$(F, <)$ admits no proper, algebraic, ordered extension field.*
(2) *$(F, <)$ is real closed.*
(3) *$F(\sqrt{-1})$ is algebraically closed and $F \neq F(\sqrt{-1})$.*

*Proof*: See, e.g. [Prestel–Delzell, 2001, 1.2.10]. □

For real closed fields, the ordering is definable (in the language of fields). Namely,

$$a < b \quad \text{iff} \quad b - a \in F^2 \setminus \{0\},$$

where $F^2 = \{x^2 \mid x \in F\}$. This means that in this (the real closed) case, $P_< = F^2$.

From (3) it follows that for $F$ real closed, an irreducible polynomial $f \in F[X]$ must be of the form

$$f = X - a \quad \text{or} \quad f = (X - b)^2 + c^2,$$

with $a, b, c \in F$ and $c \neq 0$. If we factor an arbitrary polynomial $f \in F[X]$ into such factors, we see that the *intermediate value property* must hold for $f$ over $(F, <)$; i.e.

($R_1$) *If $a, b \in F$ with $a < b$, and if $f(a) < 0 < f(b)$, then there is a $c \in F$ with $a < c < b$ and $f(c) = 0$.*

From the definition of a real closed field we immediately see the correctness of:

($R_2$) *The relative algebraic closure of a subfield in a real closed field is itself again real closed.*

As we had explained above, to every ordered field $(F, <)$, there is a real closed, algebraic extension $(F^*, <^*)$. Such an extension is called a *real closure* of $(F, <)$.

**Theorem 4.2.2.** *Any two real closures of an ordered field $(F, <)$ are isomorphic over $F$.* □

After these preparations, we are in a position to prove the main theorem of this section.

**Theorem 4.2.3.** *The theory of real closed fields admits quantifier elimination, and is complete and model complete.*

*Proof*: Model completeness follows immediately from quantifier elimination; and completeness follows from Corollary 3.4.2, since $\mathbb{Q}$ with its unique ordering is a prime substructure.

For the proof of quantifier elimination, we again proceed as in Theorems 3.4.4 and 4.1.1. Let $F_1 = (F_1, <_1)$ and $F_2 = (F_2, <_2)$ be real closed fields, and $A$ a common substructure. $A$ is, then, an ordered subring of $F_1$ and $F_2$. The construction of fractions of $A$ in $F_1$ and $F_2$ leads to canonically isomorphic ordered fields – the field of fractions $F$ of $A$. By ($R_2$) and Theorem 4.2.2, the relative algebraic closures of $F$ in $F_1$ and $F_2$ are again real closed and isomorphic over $F$. We identify these closures, and denote them by $\widetilde{F}$. Then we have the following situation:

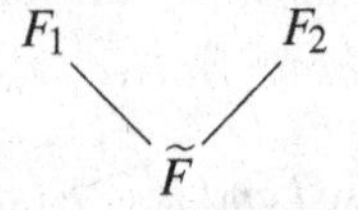

where all three structures are real closed (ordered) fields.

As in Theorem 4.1.1, we again consider a quantifier-free formula $\delta$ of the language $L(A)$ with $F_1 \models \exists x\, \delta$, in which at most $x$ occurs free. By passing to a disjunctive normal form and replacing negated atomic formulae as in the proof of Theorem 4.1.1, we may assume, after distributing the existential quantifiers to the disjuncts, that $\delta$ is of the form

$$\left( \bigwedge_{i=1}^{r} p_i(x) \doteq 0 \wedge \bigwedge_{j=1}^{s} 0 < q_j(x) \right),$$

where $p_i$ and $q_j$ are polynomials in $x$ with coefficients of the form $\underline{a}$ with $a \in A$. Here we have again carried out several field-theoretically equivalent transformations; e.g. we have gone from $p'(x) \doteq p''(x)$ to $p'(x) - p''(x) \doteq 0$. The cases $r = 0$ and $s = 0$ are not excluded.

Now, if at least one equation $p_i(x) \doteq 0$ is nontrivial, then an element $d \in F_1$ that satisfies $\delta$ in $F_1$ obviously must lie in $\widetilde{F}$. We therefore have $\widetilde{F} \models \exists x\, \delta$, and hence $F_2 \models \exists x\, \delta$. Now suppose all equations $p_i(x) \doteq 0$ are trivial, or $r = 0$. Under the assumption that there is an element $d \in F_1$ that satisfies all the inequalities $0 < q_j(x)$ in $F_1$, we shall show that $d$ can already be chosen from $\widetilde{F}$. Then we shall have $\widetilde{F} \models \exists x\, \delta$ and hence $F_2 \models \exists x\, \delta$.

Let $a_1 < \cdots < a_m$ be all the zeros in $\widetilde{F}$ of the polynomials $q_1, \ldots, q_s$, ordered according to the ordering on $\widetilde{F}$. Since the coefficients of these polynomials lie in $A$, these are also exactly their zeros in $F_1$ and $F_2$, respectively. For $d \in F_1 \setminus A$ there are then three possibilities for its location with respect to $a_1, \ldots, a_m$:

*Case 1*: $d < a_1$. In this case, $q_j(d)$ and $q_j(a_1 - 1)$ have the same sign. Namely, if they had different signs, then by ($R_1$), some zero of $q_j$ would lie between $d$ and $a_1 - 1$. But this is impossible, since the $a_\nu$ are all the zeros in $F_1$ of all the $q_j$.

*Case 2*: $a_m < d$. Here the analogous argument works for $d$ and $a_m + 1$.

*Case 3*: $a_\nu < d < a_{\nu+1}$. Here $q_j(d)$ and $q_j(\frac{a_\nu + a_{\nu+1}}{2})$ obviously again have the same sign.

In all three cases we therefore find a $d' \in \widetilde{F}$ that satisfies all inequalities $0 < q_j(x)$ in $\widetilde{F}$ just as $d$ does. □

**Corollary 4.2.4.** *The theory of ordered fields has a model companion, namely, the theory of real closed fields.*

This follows immediately from the existence of the real closure of an ordered field and from Theorem 4.2.3.

Just as for algebraically closed fields, we obtain from quantifier elimination for real closed fields $F$ that the set of quantifier-free definable sets is closed under projection. In real algebraic geometry, such sets sets are called *semi-algebraic*. We therefore have:

*The projections of semi-algebraic sets in $F^n$ are again semi-algebraic.*

The completeness of the theory of real closed fields means exactly that every sentence in the language of ordered fields that is true in $\mathbb{R}$ holds also in every other real closed field. This *transfer principle*, named after *Tarski*, finds many applications in real algebraic geometry.

One application of model completeness is the following solution of *Hilbert's 17th problem*. This problem was solved by E. Artin in 1926. His solution probably inspired certain model theoretic concepts.

**Theorem 4.2.5** (Artin's solution to Hilbert's 17th problem). *Let $F$ be a real closed field (e.g. $\mathbb{R}$), and $f \in F[X_1,\ldots,X_n]$ a positive semidefinite polynomial, i.e. $f(a_1,\ldots,a_n) \geq 0$ for all $a_1,\ldots,a_n \in F$. Then $f$ is a sum of squares of rational functions in $x_1,\ldots,x_n$ over $F$.*

*Proof*: By a theorem of Artin and Schreier, in every field of characteristic not 2, the sums of squares are exactly those elements that are positive with respect to every ordering of the field. If $f$ were not a sum of squares in the field $F(X_1,\ldots,X_n) = F(\overline{X})$ of rational functions in $X_1,\ldots,X_n$, then by this theorem there would be an ordering $<$ of $F(\overline{X})$ such that $f < 0$. To $F(\overline{X})$ with this ordering there is, by our remark before Theorem 4.2.1, a real closure, which we shall denote by $F'$. We denote the unique ordering of $F'$ by $<'$. Then in $F'$, $f <' 0$ still holds. Therefore, the existential sentence

$$\varphi : \quad \exists x_1,\ldots,x_n \, \underline{f}(x_1,\ldots,x_n) < 0$$

holds in $F'$, where we have obtained $\underline{f}$ from $f$ by replacing all coefficients $b$ by constants $\underline{b}$. $F$ is a substructure of $F'$. By model completeness, we have $F \preceq F'$. Therefore $\varphi$ holds also in $F$. But this means that there are elements $a_1,\ldots,a_n \in F$ with $f(a_1,\ldots,a_n) < 0$. This, however, contradicts our hypothesis on $f$. Therefore $f$ is a sum of squares in $F(\overline{X})$, after all. □

We have axiomatized the theory of real closed fields in a language with a symbol $<$ for the ordering. It is possible, however, to eliminate this symbol via the equivalence

$$x < y \quad \leftrightarrow \quad \exists z\,(y - x \doteq z^2 \wedge z \neq 0), \qquad (4.2.5.1)$$

which is valid in real closed fields. We can characterize those fields possessing an ordering with respect to which they are real closed, as follows: we take the field axioms $K_1$–$K_9$ and $RK_{2n}$ for all $n \geq 1$, as well as the axioms

$RK_3: \quad \forall x,y\, \exists z \qquad x^2 + y^2 \doteq z^2$

$RK_5: \quad \forall x \qquad -1 \neq x^2$

$RK_7: \quad \forall x\, \exists y\, (x \doteq y^2 \vee -x \doteq y^2)$

(which assert that the squares form a positive cone). For such fields we can always define, via (4.2.5.1), an ordering satisfying $RK_1$. Conversely, real closed fields always satisfy these axioms; i.e. the class of fields axiomatized above consists exactly of the restrictions of real closed fields to the field language.

**Theorem 4.2.6.** *The axiom system*

$$\Sigma = \{K_1, \ldots, K_9\} \cup \{RK_1, RK_3, RK_5, RK_7\} \cup \{RK_{2n} \mid n \geq 1\}$$

*is model complete and complete, but does not admit quantifier elimination.*

*Proof*: Let $F_1$ and $F_2$ be models of $\Sigma$ with $F_1 \subseteq F_2$. Then, after the addition of the order symbol via (4.2.5.1), $\left(F_1, <^{F_1}\right) \subseteq \left(F_2, <^{F_2}\right)$ also holds, because a positive element of $F_1$ is a square in $F_1$, and hence remains positive also in $F_2$. From the linearity of $<^{F_1}$, however, it follows that for $a, b \in F_1$:

$$a <^{F_1} b \quad \text{iff} \quad a <^{F_2} b.$$

The structures $\left(F_\nu, <^{F_\nu}\right)$ $(\nu = 1, 2)$ are models of the theory of real closed fields. By the model completeness of this theory it follows that

$$\left(F_1, <^{F_1}\right) \preceq \left(F_2, <^{F_2}\right),$$

which trivially implies $F_1 \preceq F_2$. By Lemma 3.3.1, $\Sigma$ is therefore model complete.

The completeness of $\Sigma$ follows similarly from the completeness of the theory of real closed fields.

Finally, we show that $\Sigma$ does not admit quantifier elimination. For this, we consider the field $\mathbb{Q}(\sqrt{2})$. This field possesses two orderings, $<_1$ and $<_2$, induced by the two embeddings of $\mathbb{Q}[X]/(X^2 - 2)$ into $\mathbb{R}$. With respect to one ordering, $\sqrt{2}$ is positive – say, $0 <_1 \sqrt{2}$, while with respect to the other, $\sqrt{2}$ is negative – say, $\sqrt{2} <_2 0$.

Let $F_1'$ and $F_2'$ be the real closures of $\left(\mathbb{Q}(\sqrt{2}), <_1\right)$ and $\left(\mathbb{Q}(\sqrt{2}), <_2\right)$, respectively. Then $F_1'$ and $F_2'$ are ordered fields, and their restrictions $F_1$ and $F_2$ to the field language are models of $\Sigma$. They contain $\mathbb{Q}(\sqrt{2})$ as a common substructure (in the field language!). However, we have

$$F_1 \models \exists x\, x^2 \doteq \underline{\sqrt{2}} \quad \text{and} \quad F_2 \models \exists x\, x^2 \doteq -\underline{\sqrt{2}}.$$

This shows, say by Theorem 3.4.1(3), that $\Sigma$ does not admit quantifier elimination. □

## 4.3 Valued Fields: Examples and Properties

The model theory of algebraic structures presented so far goes back basically to theorems and developments from the 1920s, even if the presentation has changed somewhat (especially due to the development of saturated structures in the 1950s). The model theory of valued fields, however, is a development of the 1970s. It began essentially with the sensational theorems of Ax-Kochen and Ershov in the middle of the 1960s. We shall treat these theorems in Section 4.6. Before that, however, we shall present two special cases, in Sections 4.4 and 4.5. The case of an algebraically closed valued field, handled already by A. Robinson, can thus be seen as a forerunner of all the other theorems. Since the general theory of valued fields and, in particular, Henselian fields, does not belong to the standard repertoire of an algebra course, we would like to give a short introduction in this section, even without being able to carry out proofs in full. For full proofs, we refer the interested reader to the appropriate literature.

A *valuation* of a field $F$ is a surjective mapping

$$v : F \to \Gamma \cup \{\infty\}$$

of $F$ into an ordered Abelian group $\Gamma$ (the *value group*) together with the symbol $\infty$, with the following properties:

(i) $v(a) = \infty$ iff $a = 0$
(ii) $v(ab) = v(a) + v(b)$
(iii) $v(a+b) \geq \min\{v(a), v(b)\}$,

for all $a, b \in F$. Here the symbol $\infty$ is greater than all elements of $\Gamma$, and satisfies the following computational rules:

$$\infty + \infty = \gamma + \infty = \infty + \gamma = \infty \quad \text{for} \quad \gamma \in \Gamma.$$

The valuation with $v(a) = 0$ for all $a \neq 0$ will be called the *trivial* valuation of $F$. Whenever we do not expressly emphasize that the trivial valuation should be considered, the word "valuation" will mean a nontrivial valuation.

Several properties of valuations are:

(iv) $v(1) = v(-1) = 0$
(v) $v(-a) = v(a)$, $v(a^{-1}) = -v(a)$
(vi) $v(a) < v(b)$ implies $v(a+b) = v(a)$.

These are derived as follows:

(iv) $v(1) = v(1 \cdot 1) = v(1) + v(1)$, whence $v(1) = 0$.
$0 = v(1) = v((-1) \cdot (-1)) = v(-1) + v(-1)$, whence $v(-1) = 0$.

(v) $v(-a) = v((-1) \cdot a) = v(-1) + v(a) = v(a)$.
$0 = v(1) = v(a \cdot a^{-1}) = v(a) + v(a^{-1})$, whence $v(a^{-1}) = -v(a)$.

(vi) If $v(a) < v(a+b)$, then we would have
$v(a) < \min\{v(-b), v(a+b)\} \leq v(-b+a+b) = v(a)$.

The most important examples of valuations are the following two:

*1. p-adic valuation*

Let $p$ be a prime number. A given rational number $r \neq 0$ can be decomposed uniquely (up to sign), as follows:

$$r = p^m \cdot \frac{n_1}{n_2} \quad \text{with} \quad n_1, n_2, m \in \mathbb{Z},\ n_2 \neq 0,$$

where $n_1$ and $n_2$ are not divisible by $p$. The $p$-adic value of $r$ is then defined as $v_p(r) = m$. Setting $v(0) = \infty$, we obtain

$$v_p : \mathbb{Q} \to \mathbb{Z} \cup \{\infty\}.$$

One may easily check that $v_p$ is a valuation. If the nonzero rational number $r$ is written (uniquely) as

$$r = a_m p^m + a_{m+1} p^{m+1} + \cdots + a_l p^l$$

with $m, l \in \mathbb{Z}$, $a_\nu \in \{0, 1, \ldots, p-1\}$ for $m \leq \nu \leq l$, and $a_m \neq 0$, then $v_p(r) = m$. This follows from (vi) and the observation

$$v(1) = \cdots = v(p-1) = 0.$$

The $p$-adic valuation defines a metric on $\mathbb{Q}$ if one sets

$$|x-y|_p = e^{-v_p(x-y)}.$$

Just as $\mathbb{Q}$ with respect to the metric $|x-y|$ given by the usual absolute value can be completed to the real numbers, so $\mathbb{Q}$ with respect to $|x-y|_p$ can be completed to the so-called $p$-adic number field $\mathbb{Q}_p$. The elements of $\mathbb{Q}_p$ are then precisely the infinite series of the form

$$\sum_{\nu=m}^{\infty} a_\nu p^\nu,$$

with $m \in \mathbb{Z}$ and $a_\nu \in \{0, 1, \ldots, p-1\}$. Such a series is nothing more than the limit of its subseries

$$\sum_{\nu=m}^{l} a_\nu p^\nu$$

as $l \to \infty$ in the $p$-adic metric. If we set

$$v_p\left(\sum_{\nu=m}^{\infty} a_\nu p^\nu\right) = m \quad \text{in case } a_m \neq 0,$$

then we obtain a canonical extension of $v_p$ to $\mathbb{Q}_p$ (which we continue to denote by $v_p$). Then

$$v_p : \mathbb{Q}_p \to \mathbb{Z} \cup \{\infty\}$$

is a valuation of $\mathbb{Q}_p$. The set

$$\mathbb{Z}_p = \{x \in \mathbb{Q}_p \mid v_p(x) \geq 0\}$$

forms a subring of $\mathbb{Q}_p$, the ring of *p-adic integers*. They can be represented by series

$$x = \sum_{\nu=0}^{\infty} a_\nu p^\nu,$$

where $a_0$ need not necessarily be $\neq 0$. The set of $p$-adic integers $x$ with $a_0 = 0$, i.e. with $v_p(x) > 0$, consists precisely of the ideal $p\mathbb{Z}_p$, and we have

$$\mathbb{Z}_p/p\mathbb{Z}_p \cong \mathbb{Z}/p\mathbb{Z}.$$

*2. Polynomial valuation*

Let $k$ be an arbitrary field, $k[X]$ the ring of polynomials in $X$, and $k(X)$ the field of rational functions in $X$. Then if $p \in k[X]$ is an irreducible (monic) polynomial, then a valuation $v_p$ can be defined analogously to the $p$-adic case (where $p$ now stands for an irreducible polynomial). Namely, each nonzero element $r$ of $k(X)$ can be decomposed uniquely (up to constant factors), as follows:

$$r = p^m \cdot \frac{f}{g} \quad \text{with} \quad m \in \mathbb{Z} \quad \text{and} \quad f, g \in k[X], \; g \neq 0,$$

where $f$ and $g$ are not divisible by $p$ in $k[X]$. Then if we set $v_p(r) = m$ and $v_p(0) = \infty$, this defines a valuation

$$v_p : k(X) \to \mathbb{Z} \cup \{\infty\}.$$

Trivially, $v_p(a) = 0$ for all $a \in k^\times$.

In the special case where $p = X - a$ with $a \in k$, if the nonzero rational function $r \in k(X)$ is written (uniquely) in the form

$$r = a_m p^m + a_{m+1} p^{m+1} + \cdots + a_l p^l$$

with $m, l \in \mathbb{Z}$, $a_i \in k$, and $a_m \neq 0$, then $v_p(r) = m$.

As in the $p$-adic case, $k(X)$ can be completed with respect to $v_p$. In the case $p = X - a$, the elements of this completion are the limit values of series of the form

$$\sum_{\nu=m}^{l} a_\nu (X-a)^\nu \quad \text{with} \quad a_\nu \in k$$

as $l \to \infty$. We also write

$$\sum_{\nu=m}^{\infty} a_\nu (X-a)^\nu$$

for these limit values, and call them *formal Laurent series*. With the usual computational rules for Laurent series, they form a field $k((X-a))$, the completion of $k(X)$ with respect to $v_{X-a}$. Observe that for different $a \in k$, these fields are isomorphic. The valuation $v_{X-a}$ can be canonically extended to $k((X-a))$, by defining

$$v_{X-a}\left(\sum_{\nu=m}^{\infty} a_\nu (X-a)^\nu\right) = m \quad \text{in case} \quad a_m \neq 0.$$

The formal power series in $X - a$, i.e. the series of the form

$$f = \sum_{\nu=0}^{\infty} a_\nu (X-a)^\nu,$$

form a subring $k[[X-a]]$ of $k((X-a))$. The set of power series $f$ with $a_0 = 0$ is the ideal $(X-a)k((X-a))$, and

$$k[[X-a]] \;/\; (X-a)k[[X-a]] \cong k.$$

The valuations $v_p$ just described are all trivial on $k$; i.e. $v_p(a) = 0$ for $a \in k^\times$. Conversely, one can easily show that a valuation

$$v : k(X) \to \mathbb{Z} \cup \{\infty\}$$

that is trivial on $k$ belongs to an irreducible polynomial $p \in k[X]$ or to the so-called *degree valuation* $v_\infty$. For a quotient $f/g$ with $f, g \in k[X]$ and $g \neq 0$, this valuation is defined via

$$v_\infty\left(\frac{f}{g}\right) = \deg g - \deg f.$$

If we again set $v_\infty(0) = \infty$, then

$$v_\infty : k(X) \to \mathbb{Z} \cup \{\infty\}$$

is a valuation that is trivial on $k$.

After these examples, we would like to study valuations in general. Let

$$v : F \to \mathbb{Z} \cup \{\infty\}$$

be a valuation of the field $F$. Then the set

$$\mathcal{O}_v = \{x \in F \mid v(x) \geq 0\}$$

obviously forms a subring of $F$ with the property:

$$x \notin \mathcal{O}_v \text{ implies } x^{-1} \in \mathcal{O}_v, \quad \text{for all } x \in F.$$

Subrings of $F$ with this property are called *valuation rings*. $\mathcal{O}_v$ is called the *valuation ring of* $v$. The valuation $v$ is obviously trivial on $F$ if and only if $\mathcal{O}_v = F$. Furthermore, we immediately see that the set

$$\mathfrak{M}_v = \{x \in F \mid v(x) > 0\}$$

forms an ideal of $\mathcal{O}_v$. More precisely, $\mathfrak{M}_v$ is the ideal of nonunits in $\mathcal{O}_v$, since the units of $\mathcal{O}_v$ are obviously those $x \in \mathcal{O}_v$ with $v(x) = 0$. It is thus clear that $\mathfrak{M}_v$ is a maximal ideal of $\mathcal{O}_v$, and, indeed, the only such. We call the field

$$\overline{K_v} := \mathcal{O}_v / \mathfrak{M}_v$$

the *residue field* of $v$. For the $p$-adic valuation $v_p$, the residue field is isomorphic to

$$\mathbb{F}_p = \mathbb{Z}/p\mathbb{Z},$$

both for $\mathbb{Q}$ and for $\mathbb{Q}_p$. In the polynomial case, for an irreducible $p \in k[X]$, the residue field is isomorphic to

$$k[X]/pk[X],$$

which is a finite extension of $k$ of degree $\deg p$. The degree valuation $v_\infty$ has residue field isomorphic to $k$.

Now let $\mathcal{O}$ be an arbitrary valuation ring of $F$. We do not consider the trivial case $\mathcal{O} = F$. We shall now associate with $\mathcal{O}$ a (non-trivial) valuation, whose valuation ring is exactly $\mathcal{O}$. Let $\mathcal{O}^\times$ be the group of units of $\mathcal{O}$. For elements $a, b \in F^\times$, we define an ordering of the cosets of $\mathcal{O}^\times$ in $F^\times$ by

$$a\mathcal{O}^\times < b\mathcal{O}^\times \quad \text{iff} \quad ba^{-1} \in \mathcal{O} \setminus \mathcal{O}^\times.$$

We write the multiplication of cosets additively:

$$a\mathcal{O}^\times + b\mathcal{O}^\times = ab\mathcal{O}^\times.$$

Since $ab^{-1} \in \mathcal{O}$ or $ba^{-1} \in \mathcal{O}$, it therefore follows that $a\mathcal{O}^\times > b\mathcal{O}^\times$ or $a\mathcal{O}^\times < b\mathcal{O}^\times$. The transitivity of the ordering follows from the that fact that $\mathcal{O} \setminus \mathcal{O}^\times$ is multiplicatively closed; the monotonicity is clear. Then for $a \in F^\times$ we set

$$v_{\mathcal{O}}(a) = a\mathcal{O}^\times$$

and $v_{\mathscr{O}}(0) = \infty$. This defines a valuation of $F$: Since (i) and (ii) are trivial, it remains to show (iii). Let $v_{\mathscr{O}}(a) \leq v_{\mathscr{O}}(b)$, i.e. $ba^{-1} \in \mathscr{O}$. From this follows $(a+b)a^{-1} = 1 + ba^{-1} \in \mathscr{O}$. Then

$$v_{\mathscr{O}}(a) \leq v_{\mathscr{O}}(a+b).$$

We further have

$$\mathscr{O}_{v_{\mathscr{O}}} = \{a \in F \mid v_{\mathscr{O}}(a) \geq 1\} = \mathscr{O}$$

(where 1 denotes the coset $1\mathscr{O}^{\times} = \mathscr{O}^{\times} \in F^{\times}/\mathscr{O}^{\times}$), and we see that $\mathscr{O} \setminus \mathscr{O}^{\times} = \{a \in F \mid v(a) > 1\}$ is the unique maximal ideal of $\mathscr{O}$.

If $v$ is given, and to $\mathscr{O}_v$ we form the valuation $w = v_{\mathscr{O}_v}$ as we just did, then naturally we cannot deduce that $\Gamma$ and $F^{\times}/\mathscr{O}_v^{\times}$ are identical. However, the following holds. The inverse image of $0 \in \Gamma$ under $v$ is the group $\mathscr{O}_v^{\times}$ of units. We further have

$$v(a) < v(b) \quad \text{iff} \quad ba^{-1} \in \mathscr{O}_v \setminus \mathscr{O}_v^{\times}.$$

We thereby obtain an order-preserving isomorphism $\sigma : F^{\times}/\mathscr{O}_v^{\times} \to \Gamma$ induced by $a \mapsto v(a)$. We have

$$v = \sigma \circ w.$$

Two valuations $v$ and $w$ that are related by such an isomorphism are called *equivalent*. We have thereby obtained a bijective correspondence between valuation rings and equivalence classes of valuations.

Now we wish to study extensions of valuations to extension fields. For this it is always better (especially when we look ahead to axiomatizability, later) to work with valuation rings. So let $\mathscr{O}_1$ be a valuation ring of $F_1$. If $F_2$ is an extension field of $F_1$, and $\mathscr{O}_2$ is a valuation ring of $F_2$, then we call $\mathscr{O}_2$ an *extension of* $\mathscr{O}_1$ if $\mathscr{O}_1 = F \cap \mathscr{O}_2$, i.e. if $(F_1, \mathscr{O}_1)$ is a substructure, in the sense of model theory, of $(F_2, \mathscr{O}_2)$, in which case we write $(F_1, \mathscr{O}_1) \subseteq (F_2, \mathscr{O}_2)$, as usual. If $\mathfrak{M}_i$ is the maximal ideal of $\mathscr{O}_i$, and $U_i = \mathscr{O}_i \setminus \mathfrak{M}_i$ the group of units of $\mathscr{O}_i$, for $i = 1, 2$, then, using $\mathscr{O}_1 = F_1 \cap \mathscr{O}_2$, it follows that

$$\mathfrak{M}_1 = F_1 \cap \mathfrak{M}_2 \quad \text{and} \quad U_1 = F_1 \cap U_2.$$

Here we use the equivalence

$$a \notin \mathscr{O}_i \quad \text{iff} \quad a^{-1} \in \mathfrak{M}_i$$

for $a \in F_i^{\times}$. We therefore obtain canonical injections of the residue class fields

$$\overline{F_1} := \mathscr{O}_1/\mathfrak{M}_1 \;\to\; \mathscr{O}_2/\mathfrak{M}_2 =: \overline{F_2}$$

and of the value groups

$$\Gamma_1 := F_1^{\times}/U_1 \;\to\; F_2^{\times}/U_2 =: \Gamma_2.$$

If we identify each with its image, then $\overline{F_1}$ is a subfield of $\overline{F_2}$, and $\Gamma_1$ is a subgroup of $\Gamma_2$ (with the inherited order). We call the degree

$$f = [\overline{F_2} : \overline{F_1}]$$

of the field extension the *residue degree*, and the index

$$e = [\Gamma_2 : \Gamma_1]$$

the *ramification index* of $(F_2, \mathscr{O}_2)$ over $(F_1, \mathscr{O}_1)$. If $e = f = 1$, we call it an *immediate extension*.

In general, a valuation ring $\mathscr{O}_1$ of $F_1$ has many extensions to an extension field $F_2$. The *valued field* $(F_1, \mathscr{O}_1)$ is called *Henselian* if to every algebraic extension field $F_2$ of $F_1$, $\mathscr{O}_1$ has exactly one extension to $F_2$. Therefore every valuation ring on every algebraically closed field $F$ is, trivially, Henselian.

Now we collect, in a series of theorems (mostly without proof), the most essential properties of valued and, in particular, Henselian, fields, which we shall need in the next three sections.

**Theorem 4.3.1.** *Let $(F, \mathscr{O})$ be a field with a valuation ring, and $F_1$ an extension field of $F$. Then:*
(1) (Chevalley Extension Theorem) *There always exists an extension $\mathscr{O}_1$ of $\mathscr{O}$ to $F_1$.*
(2) *If $F_1$ has degree $n$ over $F$, then for every extension $\mathscr{O}_1$ of $\mathscr{O}$ to $F_1$, the "fundamental inequality" $e \cdot f \leq n$ holds.*

*Proof*: See [Engler–Prestel, 2005, (3.1.1) and (3.2.3)]. □

If $(F, \mathscr{O})$ is a valued field and $a \in \mathscr{O}$, then we write $\overline{a}$ for the residue class $a + \mathfrak{M}$ of $a$ with respect to the maximal ideal $\mathfrak{M}$ of $\mathscr{O}$. We usually denote the residue class field $\mathscr{O}/\mathfrak{M}$ briefly by $\overline{F}$. If $f \in \mathscr{O}[X]$ is a polynomial with coefficients in $\mathscr{O}$ – say,

$$f = a_n X^n + \cdots + a_0 \quad \text{with} \quad a_i \in \mathscr{O}$$

– then

$$\overline{f} := \overline{a_n} X^n + \cdots + \overline{a_0}$$

is a polynomial over $\overline{F}$.

**Theorem 4.3.2.** *For a valued field $(F, \mathscr{O})$, the following are equivalent:*
(1) *$(F, \mathscr{O})$ is Henselian.*
(2) *If $f, g, h \in \mathscr{O}[X]$ are monic polynomials with $\overline{f} = \overline{g} \cdot \overline{h}$, and if $\overline{g}$ and $\overline{h}$ are relatively prime in $\overline{F}[X]$, then there are monic polynomials $g_1, h_1 \in \mathscr{O}[X]$ with $f = g_1 \cdot h_1$ and $\overline{g} = \overline{g_1}$, $\overline{h} = \overline{h_1}$.*
(3) *If $f \in \mathscr{O}[X]$ is monic, and if $a \in \mathscr{O}$ is a simple zero of $\overline{f}$ (i.e. $\overline{f}(\overline{a}) = \overline{0}$ and $\overline{f}'(\overline{a}) \neq \overline{0}$), then there is $a_1 \in \mathscr{O}$ with $f(a_1) = 0$ and $\overline{a_1} = \overline{a}$.*

*Proof*: See [Engler–Prestel, 2005, 4.1.3]. □

**Theorem 4.3.3.** *For each valued field $(F, \mathscr{O})$, there is a Henselian extension $(F_1, \mathscr{O}_1)$ that embeds, uniquely over $(F, \mathscr{O})$, in every other Henselian extension $(F_2, \mathscr{O}_2)$ of $(F, \mathscr{O})$.*

*Proof*: See [Engler–Prestel, 2005, 5.2.2]. □

Here, the phrase "embeds uniquely over $(F, \mathcal{O})$" means, as usual, that there is a uniquely determined monomorphism $\sigma : (F_1, \mathcal{O}_1) \to (F_2, \mathcal{O}_2)$ with $\sigma|_F = \mathrm{id}$. The (up to isomorphism uniquely determined) Henselian extension $(F_1, \mathcal{O}_1)$ given by Theorem 4.3.3 is called the *Henselization* of $(F, \mathcal{O})$. One can prove – say, with the help of condition (2) of Theorem 4.3.2 – that the relative algebraic separable closure $F'$ of a field $F$ in an extension field $F_1$ with respect to a valuation $\mathcal{O}' = F' \cap \mathcal{O}_1$ is Henselian, in case $(F_1, \mathcal{O}_1)$ is a Henselian field. From this we see that the Henselization of a valued field is always a separable extension. Moreover, we have:

**Theorem 4.3.4.** *The Henselization $(F_1, \mathcal{O}_1)$ of a valued field $(F, \mathcal{O})$ is a separable and immediate extension of $(F, \mathcal{O})$.* □

We shall call a nontrivial valued field $(F, \mathcal{O})$ *finitely ramified* if either $\operatorname{char} \overline{F} = 0$ or only finitely many values in the value group $\Gamma$ lie between 0 and $v(\operatorname{char} \overline{F})$. From the second case it follows, as one may easily check, that for every natural number $n \geq 1$, only finitely many values in $\Gamma$ lie between 0 and $v(n)$. In both cases, $\operatorname{char} F$ is necessarily 0.

**Theorem 4.3.5.** *Let $(F, \mathcal{O})$ be a finitely ramified valued field. Then the Henselization of $(F, \mathcal{O})$ is a maximal immediate algebraic extension.*

*Proof*: By Theorem 4.3.4, the Henselization $(F_1, \mathcal{O}_1)$ of $(F, \mathcal{O})$ is an immediate extension. Assume $(F_2, \mathcal{O}_2)$ is a proper, immediate, algebraic extension of $(F_1, \mathcal{O}_1)$. Let, say, $F_2 = F_1(\alpha)$, $g = \operatorname{Irr}(\alpha, F_1)$, and $F_3$ be the splitting field of $g$ over $F_1$. Then $F_3/F_1$ is a Galois extension, whose Galois group $G$ has, say, $n > 1$ elements. The sum

$$a = \frac{1}{n} \sum_{\sigma \in G} \sigma(\alpha)$$

lies in $F_1$. We shall show the existence of a $b \in F_1$ with

$$v(\alpha - a) + v(n) < v(\alpha - b). \tag{4.3.5.1}$$

Since $(F_2, \mathcal{O}_2)$ is an immediate extension of $(F_1, \mathcal{O}_1)$, there is, first, a $c \in F_1^\times$ with $v(\alpha - a) = v(c)$. Then $(\alpha - a)c^{-1}$ is a unit in $\mathcal{O}_2$, whence there is, secondly, a $d \in \mathcal{O}_1$ with $\overline{(\alpha - a)c^{-1}} = \overline{d}$. If we write $b_1 = a + cd$, it follows that

$$v(\alpha - a) = v(c) < v\Big(c \cdot \Big(\frac{\alpha - a}{c} - d\Big)\Big) = v(\alpha - b_1).$$

Since only finitely many values lie between $v(\alpha - a)$ and $v(\alpha - a) + v(n)$, we can, after finitely many applications of this process, find a $b \in F_1$ satisfying (4.3.5.1).

Since all automorphisms of $F_3$ over $F_1$ must carry the unique extension $\mathcal{O}_3$ of $\mathcal{O}_1$ (or of $\mathcal{O}_2$), into itself (just because of its uniqueness), it follows that

$$v(\sigma(\alpha) - b) = v(\alpha - b)$$

for all $\sigma \in G$. We therefore obtain

$$v(\alpha - a) + v(n) < v(\alpha - b) \leq v\Bigg(\sum_{\sigma \in G} (\sigma(\alpha) - b)\Bigg).$$

Since $\sum_{\sigma \in G}(\sigma(\alpha) - b) = n(a - b)$, it thus follows that

$$v(\alpha - a) < v(a - b).$$

But this and (4.3.5.1) yield

$$v(\alpha - a) < \min\{v(a - b),\, v(\alpha - b)\} \leq v((\alpha - b) - (a - b)),$$

a contradiction.

Thus, $(F_1, \mathscr{O}_1)$ is a maximal immediate algebraic extension of $(F, \mathscr{O})$. □

## 4.4 Algebraically Closed Valued Fields

Now we would like to study the model theory of algebraic and real closed fields with valuation rings. Here we do not yet use the theorems on Henselian fields.

For the first case we use the language of fields, in the second case the language of ordered fields (as introduced in Section 1.6), which, however, we extend in both cases by a 1-place relation symbol $V$. The following axioms express that the interpretation of $V$ is a valuation ring (in a field):

$$\begin{aligned} &\mathrm{V}_1: \quad V(0) \wedge V(1) \\ &\mathrm{V}_2: \quad \forall x, y\, (V(x) \wedge V(y) \rightarrow V(x - y) \wedge V(x \cdot y)) \\ &\mathrm{V}_3: \quad \forall x, y\, (x \cdot y = 1 \rightarrow V(x) \vee V(y)). \end{aligned}$$

We write a model of the field axioms $\mathrm{K}_1$–$\mathrm{K}_9$ and $\mathrm{V}_1$–$\mathrm{V}_3$ briefly as $(F, \mathscr{O})$, where $F$ is a field and $\mathscr{O}$ is the interpretation of $V$, hence a valuation ring of $F$. First we consider the case where $K$ is algebraically closed, so that the axioms $\mathrm{AK}_n$ for $n \geq 1$ are also fulfilled.

In order to get results on quantifier elimination, it is necessary to formulate the concept of valuation so that it reappears in substructures, in a certain sense. But substructures of fields are integral domains. We therefore make the following modification: let $F$ be a field and $\mathscr{O}$ a valuation ring of $F$. For $a, b \in F$ we define a divisibility relation by

$$a \mid b \quad \text{iff} \quad ac = b \text{ for some } c \in \mathscr{O}.$$

We see immediately that $\mathscr{O}$ consists exactly of those $a \in F$ that are divisible by 1. For $a, b \in F$ we also have:

$$a \mid b \quad \text{iff} \quad v_{\mathscr{O}}(a) \leq v_{\mathscr{O}}(b).$$

Soon we shall express the following properties of divisibility in the form of axioms. Here the symbol $|$ is a binary relation symbol, which extends the language of fields.

$$\begin{aligned}
&D_1: && (1 \mid 0 \wedge 0 \nmid 1)\\
&D_2: && \forall x \quad x \mid x\\
&D_3: && \forall x,y,z\,(x \mid y \wedge y \mid z \rightarrow x \mid z)\\
&D_4: && \forall x,y \quad (x \mid y \vee y \mid x)\\
&D_5: && \forall x,y,z\,(x \mid y \rightarrow xz \mid yz)\\
&D_6: && \forall x,y,z\,(z \mid x \wedge z \mid y \rightarrow z \mid x+y)
\end{aligned}$$

Now if $F$ is a field with a binary relation (which we continue to denote by $|$) that satisfies $D_1$–$D_6$, then

$$\mathscr{O} := \{a \in F \mid 1 \mid a\}$$

defines a valuation ring. One sees this easily. (We get $-1 \in \mathscr{O}$ from $1 \mid -1 \vee -1 \mid 1$, using $D_5$.) In the same way we see that we have thereby produced, in any field, a one-to-one correspondence between valuation rings and divisibility relations satisfying $D_1$–$D_6$. We therefore call such a divisibility relation a *valuation divisibility*. The advantage of valuation divisibility lies in the following

**Lemma 4.4.1.** *Let $R$ be an integral domain with a divisibility relation $|$ satisfying* $D_1$–$D_6$. *Then there is a unique extension of $|$ to the quotient field $F$ of $R$ that continues to satisfy* $D_1$–$D_6$.

*Proof*: For $a,b,c,d \in R$ with $b,d \neq 0$, we define

$$\frac{a}{b} \,\Big|\, \frac{c}{d} \quad \text{iff} \quad ad \mid bc. \tag{4.4.1.1}$$

It is easy to convince oneself that this relation on $F$ is well-defined, extends the given $|$ on $R$, and satisfies the axioms $D_1$–$D_6$. On the other hand, every extension satisfying $D_1$–$D_6$ also satisfies (4.4.1.1), and so is uniquely determined by the divisibility relation $|$ on $R$. □

A valuation divisibility $|$ corresponds to the trivial valuation ring of a field $F$ if and only if 1 divides every element of $F$.

**Theorem 4.4.2.** *The theory of algebraically closed fields with nontrivial valuation divisibility admits quantifier elimination.*

*Proof*: We use criterion (3) of Theorem 3.4.1. Accordingly, let $(F_1, |_1)$ and $(F_2, |_2)$ be algebraically closed fields with nontrivial valuation divisibilities, and let $(F, |)$ be a common substructure. Using Lemma 4.4.1, we may assume that $F$ is already a field. If we take the algebraic closures of $F$ in $F_1$ and in $F_2$, these are naturally isomorphic, as fields, over $F$. There is also, however, an isomorphism that carries the valuation ring induced by $|_1$ over to that induced by $|_2$. This follows from the fact that two valuation rings $\mathcal{O}_1$ and $\mathcal{O}_2$ of the algebraic closure $\widetilde{F}$ of a field $F$ that both extend the valuation ring $\mathcal{O}$ of $F$ can always be mapped to each other by an automorphism of $\widetilde{F}$ over $F$. (Cf. [Engler–Prestel, 2005, 3.2.15].) We may therefore assume, finally, that $F$ itself is algebraically closed.

We must now show that an arbitrary, simple existence sentence $\varphi$ of the language $L(F, |\ )$ that holds in $(F_1, |_1)$, also holds in $(F_2, |_2)$. The sentence $\varphi$ obviously holds already in a simple, pure transcendental extension $F' = F(t)$ of $F$. We shall show that, with respect to the valuation ring $\mathcal{O}'$ in $F'$ induced by $|_1$, $F'$ embeds in $F_2$ over $F$. Then $\varphi$ will hold also in $(F_2, |_2)$. In this proof we shall also assume that $(F_2, |_2)$ is $\kappa^+$-saturated, where $\kappa = \text{card}(F)$. This is no restriction, since we can at any time pass to a corresponding saturated elementary extension of $(F_2, |_2)$, using Theorem 2.5.2.

Let $\mathcal{O}_2$ be the valuation ring of $F_2$ induced by $|_2$. We seek a monomorphism

$$\sigma : (F', \mathcal{O}') \to (F_2, \mathcal{O}_2)$$

with $\sigma|_F = \text{id}$. We shall distinguish three cases. Here, $\overline{F}, \overline{F'}, \overline{F_2}$ will always denote the residue fields of $(F, \mathcal{O})$, $(F', \mathcal{O}')$ and $(F_2, \mathcal{O}_2)$, respectively, and $\Gamma, \Gamma', \Gamma_2$ the corresponding value groups.

*Case 1*: $\overline{F'} \neq \overline{F}$. Let, say, $\overline{x} \in \overline{F'} \setminus \overline{F}$. Then $\overline{x}$ is transcendental over $\overline{F}$, since $\overline{F}$ is algebraically closed. This is proved as follows: if $\overline{x}$ were algebraic over $\overline{F}$, and if, say, $\overline{g}$ with

$$g = X^m + a_{m-1}X^{m-1} + \cdots + a_0 \in \mathcal{O}[X]$$

were the minimal polynomial of $\overline{x}$ over $\overline{F}$, then every factorization of $g$ would carry over to $\overline{g}$. Since $F$ is algebraically closed, $g$ would have to be linear.

Now we consider an inverse image $x$ of $\overline{x}$. By Theorem 4.3.1(2), $x$ is also transcendental over $F$. We further choose an $x_2 \in F_2$ such that $\overline{x_2}$ is transcendental over $\overline{F}$. This is possible, since $(F_2, |_2)$ is $\kappa^+$-saturated. Namely, this results from the fact that the type

$$\Phi(v_0) = \{(1 \mid v_0 \wedge v_0 \mid 1)\} \cup \{(1 \mid (v_0 - \underline{a}) \wedge (v_0 - \underline{a}) \mid 1) \mid a \in \mathcal{O}^\times\}$$

is realizable in $(F_2, |_2)$. (Observe that, due to the infinitude of $\overline{F}$, the set $\Phi(v_0)$ is finitely satisfiable in $(F_2, |_2)$.) A realization $x_2$ of this type is, however, just a unit in $\mathcal{O}_2$, for which $\overline{x_2} - \overline{a} \neq \overline{0}$ for all $a \in \mathcal{O}^\times$ holds. As before, $x_2$ is then transcendental over $F$. Therefore, the mapping $a \mapsto a$ for $a \in F$ and $x \mapsto x_2$ defines a field isomorphism $\sigma$ of $F(x)$ onto $F(x_2)$. This even respects the valuation, since, for the valuation $v'$ induced by $\mathcal{O}'$ and for $a_0, \ldots, a_m \in F$, we have:

$$v'(a_m x^m + \cdots + a_0) = \min_{0 \leq i \leq m} v(a_i). \tag{4.4.2.1}$$

Namely, if not all $a_i$ are 0, and if, say, $v(a_j)$ is minimal, then (4.4.2.1) follows from the equation

$$v'\left(\sum_{i=0}^{m} a_i x^i\right) = v(a_j) + v'\left(\sum_{i=0}^{m} a_i a_j^{-1} x^i\right)$$

and the fact that $\overline{\sum_{i=0}^{m} a_i a_j^{-1} \overline{x}^i} \neq 0$. Here we use the facts that $\overline{x}$ is transcendental over $\overline{F}$ and that all coefficients $a_i a_j^{-1}$ lie in $\mathscr{O}$. Since the same argument works for the valuation induced on $F(x_2)$ by $\mathscr{O}_2$, we therefore obtain, finally, for $a_0, \ldots, a_m \in F$:

$$v'(a_m x^m + \cdots + a_0) = v_2(a_m x^m + \cdots + a_0).$$

We therefore have an isomorphism $\sigma$ from the subfield $F(x)$ of $F'$ onto a subfield $F''$ of $F_2$. Since $F'$ is algebraic over $F(x)$, and $F_2$ is algebraically closed, $\sigma$ can naturally be extended to $F'$. Here we recall the fact, mentioned above, that two extensions of a valuation ring of the field $F''$ to its algebraic closure $\widetilde{F''}$ can always be carried to each other by an automorphism of $\widetilde{F''}$ over $F''$.

*Case 2*: $\Gamma' \neq \Gamma$. Let, say, $v'(x) \in \Gamma' \setminus \Gamma$, for $x \in F'$. The group $\Gamma$ is divisible. This results as follows: a typical element of $\Gamma$ is $v(a)$, for $a \in F$. For $n \geq 2$, there is, then, a $b \in F$ with $a = b^n$. It therefore follows that $v(a) = nv(b)$, i.e. $v(a)$ is divisible by $n$ in $\Gamma$. Using Theorem 4.3.1(2), it follows again that $x$ must be transcendental over $F$.

Now we choose an $x_2 \in F_2$ such that $v_2(x_2)$ occupies the same location with respect to $\Gamma$ as $v'(x)$. Such an $x_2$ exists, again because of the $\kappa^+$-saturation of $(F_2, |_2)$ and the divisibility of $\Gamma$. Namely, this results from the fact that the type

$$\{\neg v_0 \mid \underline{a} \mid v(a) < v'(x),\ a \in F\} \cup \{\neg \underline{a} \mid v_0 \mid v'(x) < v(a),\ a \in F\}$$

is realizable in $(F_2, |_2)$. (In the case that $\mathscr{O}$ is trivial, we use the hypothesis that $\mathscr{O}_2$ is nontrivial.) In particular, $x_2$ is again transcendental over $F$. Therefore, the mapping $a \mapsto a$ for $a \in F$ and $x \mapsto x_2$ defines a field isomorphism of $F(x)$ with $F(x_2)$. This again respects the corresponding valuations. Namely, for $a_0, \ldots, a_m \in F$ we have

$$v'(a_m x^m + \cdots + a_0) = \min_{0 \leq i \leq m} (v(a_i) + iv'(x)). \tag{4.4.2.2}$$

This follows as in Case 1: indeed, if not all $a_i$ are 0, and if $v'(a_i x^i) = v'(a_j x^j)$ for some $i < j$, then we would have

$$(j-i)v'(x) = v(a_i a_j^{-1}) \in \Gamma.$$

For $x_2$ we obtain, analogously,

$$v_2(a_m x^m + \cdots + a_0) = \min_{0 \leq i \leq m} (v(a_i) + iv_2(x)).$$

Because of the divisibility of $\Gamma$ and the equal location with respect to $\Gamma$ occupied by $x$ and $x_2$, the mapping $\gamma \mapsto \gamma$ for $\gamma \in \Gamma$ and $v'(x) \mapsto v_2(x)$ defines an order-isomorphism of the value groups $\Gamma \oplus \mathbb{Z}v'(x)$ of $F(x)$ and $\Gamma \oplus \mathbb{Z}v_2(x_2)$ of $F(x_2)$.

This shows that $\sigma$ carries the valuation induced by $\mathcal{O}'$ on $F(x)$ to the valuation induced by $\mathcal{O}_2$ on $F(x_2)$. As in Case 1, $\sigma$ can be extended from $F(x)$ to $F'$.

*Case 3*: $\overline{F'} = \overline{F}$ and $\Gamma' = \Gamma$. We shall prove the existence of a $t_2 \in F_2$ with

$$v'(t-a) = v_2(t_2-a) \quad \text{for all } a \in F. \tag{4.4.2.3}$$

Obviously $t_2$ will then be transcendental over $F$ (since $v'(t-a) \neq \infty$ for $a \in F$). Therefore the mapping $a \mapsto a$ for $a \in F$ and $t \mapsto t_2$ defines a monomorphism of $F(t)$ into $F_2$, which respects the valuations, by (4.4.2.3) (and the fact that every polynomial in $t$ over $F$ decomposes into $F$-linear factors).

In order to prove the existence of a $t_2 \in F_2$ satisfying (4.4.2.3), we shall again apply the $\kappa^+$-saturation of $(F_2, |_2)$. We shall show that the following set of formulae is a type of $(F_2, |_2)$, i.e. is finitely satisfiable in $(F_2, |_2)$:

$$\Phi(v_0) = \{ (v_0 - \underline{a} \mid \underline{b_a} \wedge \underline{b_a} \mid v_0 - \underline{a}) \mid a \in F \}. \tag{4.4.2.4}$$

Here $b_a \in F$ is chosen so that

$$v(b_a) = v'(t-a), \tag{4.4.2.5}$$

using $\Gamma' = \Gamma$. If this type is realized by a $t_2 \in F_2$, then obviously

$$v_2(t_2-a) = v(b_a) = v'(t-a) \quad \text{for all } a \in F, \tag{4.4.2.6}$$

proving (4.4.2.3).

So let a finite subset of $\Phi(v_0)$ be given, i.e. let $a_1, \dots, a_n \in F$ and $b_{a_1}, \dots, b_{a_n}$ satisfying (4.4.2.5) be given. We seek an element $d \in F_2$ with

$$v_2(d-a_1) = v(b_{a_1}), \ \dots, \ v_2(d-a_n) = v(b_{a_n}).$$

For this, let $v(b_a)$ be the maximum of the $v(b_{a_i})$ for $1 \leq i \leq n$. Since $v'(t-a) = v(b_a)$, $(t-a)b_a^{-1}$ is (a unit) in $\mathcal{O}'$. Using $\overline{F'} = \overline{F}$, there is, therefore, a $c \in F$ with $\overline{c} = \overline{(t-a)b_a^{-1}}$, i.e.

$$v'\left(\frac{t-(a+cb_a)}{b_a}\right) = v'\left(\frac{t-a}{b_a} - c\right) > 0.$$

If we set $d = a + cb_a$, it therefore follows that

$$v'(t-d) > v(b_a) \geq v(b_{a_i}) = v'(t-a_i)$$

for $1 \leq i \leq n$. In particular, we then have

$$v(d-a_i) = v'((t-a_i)-(t-d)) = v'(t-a_i) = v(b_{a_i}).$$

Case 3 is hence also brought to a close. □

For the theory of nontrivial valued, algebraically closed fields – axiomatized via $K_1$–$K_9$, $AK_n$ for $n \geq 1$, $V_1$–$V_3$, and $\exists x \neg V(x)$ – we obtain from Theorem 4.4.2 the following

**Corollary 4.4.3.** *The theory of nontrivial valued, algebraically closed fields is model complete; it is the model companion of the theory of valued fields.*

*Proof*: Since the model completeness is implied trivially by the quantifier elimination, it suffices, by Theorem 4.4.2, to translate a formula from the language with the unary relation symbol $V$ into a formula with the binary relation $|$. But this is trivial: we replace $V(t)$ by $1 \mid t$ for each term $t$.

It remains to show that every valued field embeds in an algebraically closed field with a nontrivial valuation. But this follows immediately from Theorem 4.3.1(1), when one bears in mind that the trivial valuation on $F$ can always be extended to a nontrivial valuation on the rational function field $F(t)$. □

To obtain theorems on completeness, one must fix the characteristic of the field and of the residue field. If $(F, \mathcal{O})$ is a valued field with residue field $\overline{F}$, and if $\operatorname{char} \overline{F} = 0$, then obviously also $\operatorname{char} F = 0$. For $\operatorname{char} \overline{F} = p > 0$, both $\operatorname{char} F = p$ and $\operatorname{char} F = 0$ are possible. If $\overline{F}$ has characteristic $p$, this can be axiomatized by

$$\neg \exists x\, (px \doteq 1 \wedge V(x)) \quad \text{respectively} \quad \neg p \mid 1.$$

If $\operatorname{char} \overline{F} = 0$, then this is axiomatized by the set

$$\{\exists x\, (px \doteq 1 \wedge V(x)) \mid p \text{ prime}\} \quad \text{respectively} \quad \{p \mid 1 \mid p \text{ prime}\}.$$

We thus obtain

**Corollary 4.4.4.** *The theory of nontrivial valued, algebraically closed fields with fixed characteristic and fixed characteristic of the residue field is complete.*

*Proof*: As in the proof of the previous corollary, we show this in the language with $|$. By Corollary 3.4.2, it suffices to find a prime substructure. For the case $\operatorname{char} F = \operatorname{char} \overline{F} = 0$, this is $\mathbb{Q}$ with the the trivial valuation divisibility. For the case $\operatorname{char} F = \operatorname{char} \overline{F} = p$, this is the field $\mathbb{F}_p$ with $p$ elements and with the trivial valuation divisibility. For the case $\operatorname{char} F = 0$ and $\operatorname{char} \overline{F} = p$, finally, this is the field $\mathbb{Q}$ with the divisibility belonging to the $p$-adic valuation $v_p$. The $p$-adic valuation is, namely, the only valuation of $\mathbb{Q}$ whose residue field has characteristic $p$. □

## 4.5 Real Closed Valued Fields

Now we come to the case of a real closed field $(F, <)$, where we wish to emphasize the ordering, as in Section 4.2. Here we recall the axioms $K_1$–$K_9$, $RK_1$, and $RK_{2n}$ for $n \geq 1$ from Section 1.6.

In an ordered field $(F,<)$ we consider a valuation ring $\mathscr{O}$ that is *convex* with respect to $<$; i.e. for $a,b \in F$, we should have that

$$0 \leq a \leq b \in \mathscr{O} \quad \text{implies} \quad a \in \mathscr{O}.$$

The reader can easily see that this is equivalent to the statement: "for all $a,b \in F$,

$$0 \leq a \leq b \quad \text{implies} \quad v_{\mathscr{O}}(b) \leq v_{\mathscr{O}}(a)\text{''}.$$

We then call a valuation divisibility *compatible with* $<$ if, for all $a,b \in F$, we have

$$0 \leq a \leq b \quad \text{implies} \quad b \mid a.$$

Then obviously the valuation divisibilities compatible with $<$ correspond to valuation rings convex with respect to $<$.

Examples of convex valuation rings are easy to obtain: let, say, $R$ be a subring with 1 of the ordered field $(F,<)$. Then the convex hull of $R$,

$$\mathscr{O} = \{a \in F \mid |a| \leq b \text{ for some } b \in R\},$$

is a convex valuation ring. Indeed, if $a \in F$ (with $0 < a$) is not in $\mathscr{O}$, then $1 < a$. Using $0 < a^{-1} < 1$, it then follows, however, that $a^{-1} \in \mathscr{O}$.

If $\mathscr{O}$ is a convex valuation ring of the ordered field $(F,<)$, then $<$ induces a canonical ordering of the residue field $\overline{F}$ (which we shall continue to denote by $<$), as follows: for $a,b \in \mathscr{O}$ with $\overline{a} \neq \overline{b}$, we declare:

$$\overline{a} < \overline{b} \quad \text{iff.} \quad a < b.$$

We merely have to show that this definition is independent of the choice of representatives of the residue classes $\overline{a}$ and $\overline{b}$. For this, let, say, $m_1, m_2 \in \mathfrak{M}$ and $b + m_2 \leq a + m_1$. But then from $a < b$,

$$0 < b - a \leq m_1 - m_2 \in \mathfrak{M}$$

would follow. But since in a convex valuation ring the maximal ideal $\mathfrak{M}$ is also convex (a fact of which readers can easily convince themselves), it follows that $b - a \in \mathfrak{M}$, i.e. $\overline{a} = \overline{b}$.

A convex valuation ring $\mathscr{O}$ in an ordered field $(F,<)$ extends uniquely (as a convex valuation ring) to the real closure $(F_1,<_1)$ of $(F,<)$. Indeed, if we form the convex hull $\mathscr{O}_1$ of $\mathscr{O}$ in $(F_1,<_1)$, then this is, on the one hand, an extension. If, on the other hand, $\mathscr{O}_1'$ is a convex extension of $\mathscr{O}$ on $(F_1,<_1)$ that is different from $\mathscr{O}_1$, then $\mathscr{O}_1 \subsetneq \mathscr{O}_1'$ must hold. But this is impossible for algebraic extensions. Indeed, if $\mathscr{O}_1 \subsetneq \mathscr{O}_1'$, then this would have to hold already in some finite extension $F_1$ of $F$. If we write $\mathfrak{M}_1$ and $\mathfrak{M}_1'$ for the maximal ideals of $\mathscr{O}_1$ and $\mathscr{O}_1'$, respectively, then from $\mathscr{O}_1 \subseteq \mathscr{O}_1'$ we immediately get $\mathfrak{M}_1' \subseteq \mathfrak{M}_1$. Using $\mathfrak{M}_1' \cap \mathscr{O} = \mathfrak{M}$, it then follows (with the usual identifications) that

$$\mathscr{O}/\mathfrak{M} \subseteq \mathscr{O}_1/\mathfrak{M}'_1 \subseteq \mathscr{O}'_1/\mathfrak{M}'_1.$$

Since the residue degree is finite by Theorem 4.3.1(2), it follows that $\mathscr{O}_1/\mathfrak{M}'_1$ must be a field. Therefore $\mathfrak{M}'_1$ is maximal. Consequently, $\mathfrak{M}'_1 = \mathfrak{M}_1$ and thus $\mathscr{O}'_1 = \mathscr{O}_1$.

After these preparations, we can prove the following theorem:

**Theorem 4.5.1.** *The theory of real closed fields with a compatible, nontrivial valuation divisibility admits quantifier elimination.*

*Proof*: We essentially follow the proof of Theorem 4.4.2. Accordingly, let $(F_1, <_1, |_1)$ and $(F_2, <_2, |_2)$ be real closed fields with compatible valuation divisibilities, and let $(F, <, |)$ be a common substructure. By Lemma 4.4.1 and the unique extendability to the real closure of the valuation determined by $|$, we can identify the real closure of $F$ in $F_1$ with that in $F_2$. We can also assume that $F$ itself is real closed. Then, as in the proof of Theorem 4.4.2, we have to show that an arbitrary, pure transcendental subextension $F' = F(t)$ of $F$ in $F_1$ embeds in $F_2$ with respect to the valuation in $F_2$ over $F$, where we may assume that $(F_2, <_2, |_2)$ is $\kappa^+$-saturated for $\kappa = \mathrm{card}(F)$. Again we distinguish three cases (with the notations of (4.4.2)).

*Case 1*: $\overline{F'} \neq \overline{F}$. Let $\overline{x} \in \overline{F'} \setminus \overline{F}$ with $\overline{0} < \overline{x}$. Since every $F$-polynomial of odd degree, and every polynomial $X^2 - a$ with $a > 0$ in $F$, has a zero in $F$, this holds also for $\overline{F}$. Since, in addition, $\overline{F'}$ is ordered, $\overline{F}$ must therefore be algebraically closed in $\overline{F'}$. Therefore $\overline{x}$ is transcendental over $\overline{F}$, and so $x$ is also transcendental over $F$.

As in the proof of (4.4.2), we now seek $x_2 \in F_2$ such that, on the one hand, $\overline{x_2}$ is transcendental over $\overline{F}$ – whence, as in (4.4.2), the mapping $a \mapsto a$ for $a \in F$ and $x \mapsto x_2$ defines a valuation-preserving embedding $\sigma$ of $F(x)$ into $F_2$ over $F$ – and, on the other hand, $x_2$ occupies the same position (relative to the elements of $F$) with respect to the ordering $<_2$, as $x$ does with respect to $<_1$. Hence, as in (4.4.2), in order to achieve this, we consider the following set of formulae:

$$\begin{aligned}\Phi(v_0) = \{&(1 \mid v_0 \wedge v_0 \mid 1)\} \cup \\ &\{(\underline{a} < v_0 \wedge 1 \mid v_0 - \underline{a} \wedge v_0 - \underline{a} \mid 1) \mid a \in \mathscr{O}, \overline{a} <_1 \overline{x}\} \cup \\ &\{(v_0 < \underline{b} \wedge 1 \mid v_0 - \underline{b} \wedge v_0 - \underline{b} \mid 1) \mid b \in \mathscr{O}, \overline{x} <_1 \overline{b}\}.\end{aligned}$$

For any $x_2 \in F_2$ satisfying $\Phi$, the residue class $\overline{x_2}$ occupies the same position relative to the elements of $\overline{F}$ that $\overline{x}$ does. It is clear that $\Phi$ is finitely satisfiable in $F_2$. By the $\kappa^+$-saturation of $(F_2, <_2, |_2)$, it is therefore clear that there is, in fact, an $x_2 \in F_2$ satisfying $\Phi$.

For such an $x_2 \in F_2$, $\overline{x_2}$ is, in particular, transcendental over $\overline{F}$, and $x_2 > 0$. Therefore $x_2$ is also transcendental over $F$. Hence the mapping $a \mapsto a$ for $a \in F$ and $x \mapsto x_2$ defines a monomorphism $\sigma$ from $F(x)$ into $F_2$ over $F$, which can be proved to be valuation-preserving by the same argument as in the proof of (4.4.2). In order to show order-preservation, we observe, first, that $x_2$ occupies the same position relative to the elements of $F$ as $x$ does. If, say, $0 < a <_1 x$ for $a \in F$, it then follows that $\overline{0} \leq \overline{a} < \overline{x}$ and thence $\overline{0} \leq \overline{a} < \overline{x_2}$. But this last implies $a <_2 x_2$. If $x <_1 b$ for $b \in F$, one deduces $x_2 <_2 b$ analogously (where in this case $v(b)$ can be $< 0$). Now since every (monic) polynomial $f(X)$ with coefficients in the real closed field $F$ can be written as the product of factors of the forms

$$X-a \quad \text{or} \quad (X-a)^2+b^2,$$

with $a,b \in F$ and $b \neq 0$, the position of $x$ (or of $x_2$) with respect to the elements of $F$ determines the sign of $f(x)$ (or of $f(x_2)$, respectively).

Since $F'$ is algebraic over $F(x)$, the monomorphism $\sigma$ on $F(x)$ has an order-preserving extension $\sigma'$ defined on $F'$. By the unique extendability of a convex valuation ring to the real closure, $\sigma'$ is also valuation-preserving.

*Case 2*: $\Gamma' \neq \Gamma$. Let, say, $v'(x) \in \Gamma' \setminus \Gamma$, with $0 <_1 x$. Since for every $n \geq 2$ we can extract an $n$th root of any positive element of $F$, the group $\Gamma$ is divisible, as in the proof of (4.4.2). Hence, as before, $x$ is transcendental over $F$.

This time we consider the following set $\Phi(v_0)$ of formulae:

$$\{\neg v_0 \mid \underline{a} \mid a \in F,\ v(a) < v'(x)\} \cup \{\neg \underline{b} \mid v_0 \mid b \in F,\ v'(x) < v(b)\}.$$

Since the group $\Gamma_2$ is divisible and, by hypothesis, nontrivial, $\Phi$ is finitely satisfiable in $F_2$. By the $\kappa^+$-saturation of $(F_2, <_2, |_2)$, there is, then, an $x_2 \in F_2$ that satisfies $\Phi$ and is, without loss of generality, positive. Since, in particular, $v_2(x_2) \notin \Gamma$ holds, $x_2$ is, as in the proof of (4.4.2), transcendental over $F$, and the mapping $a \mapsto a$ for $a \in F$ and $x \mapsto x_2$ determines a valuation-preserving monomorphism $\sigma$ of $F(x)$ into $F_2$ over $F$. This is also order-preserving, however, since from $0 \leq a <_1 x$ or $x <_1 b$ with $a, b \in F$, it follows immediately that $v(a) > v'(x)$ or $v'(x) > v(b)$, respectively. But this has as consequences that $v(a) > v_2(x_2)$ or $v_2(x_2) > v(b)$, and thus $a <_2 x_2$ or $x_2 <_2 b$, respectively. Now the order-preservation of $\sigma$ follows as in Case 1.

Just as in Case 1, $\sigma$ can be extended to the algebraic extension $F'$ of $F(x)$.

*Case 3*: $\overline{F'} = \overline{F}$ and $\Gamma' = \Gamma$. Here, for any $t \in F_1 \setminus F$ with $t >_1 0$, we can, as in the proof of (4.4.2), find a $t_2 \in F_2$ satisfying (4.4.2.3). It remains to show that the mapping $a \mapsto a$ for $a \in F$ and $t \mapsto t_2$ defines a monomorphism $\sigma$ of $F(t)$ into $F_2$ over $F$ that preserves the orderings and the valuations. That $\sigma$ is value-preserving is just the statement of Lemma 4.5.2 below (observing also Remark 4.5.3(2) thereafter). To show that $\sigma$ is order-preserving, it again suffices to show that $t_2$ occupies the same place with respect to the elements of $F$ that $t$ does. So suppose $0 < a <_1 t$ with $a \in F$. We choose $b \in F$ with $0 < b$ and $v(b) = v'(t-a)$. Then $(t-a)b^{-1}$ is $>_1 0$ and a unit in $\mathscr{O}'$. Therefore in $\overline{F'}$ we have

$$\overline{0} < \overline{(t-a)b^{-1}} \in \overline{F}.$$

Since $\sigma$ is value-preserving, however, $\overline{(t-a)b^{-1}} = \overline{(t_2-a)b^{-1}}$. We thereby obtain, finally, $0 <_2 (t_2-a)b^{-1}$, and, using $0 < b$, also $0 <_2 t_2 - a$. For the case $t <_1 b$ with $b \in F$ we deduce $t_2 <_2 b$, analogously. □

It remains to prove the following lemma:

**Lemma 4.5.2.** *Let $(F, \mathscr{O})$ be a valued field admitting no proper, immediate, algebraic extension. Further, let $v'$ be an immediate extension of $v = v_{\mathscr{O}}$ to the rational* *function field $F' = F(t)$. Then the values $v'(t-a)$ for $a \in F$ determine $v'$ uniquely. More precisely, if $v''$ is any extension of $v$ to $F'$ satisfying*

$$v'(t-a) = v''(t-a) \quad \textit{for all } a \in F, \tag{4.5.2.1}$$

*then for every monic polynomial $f(t) \in F[t]$:*

$$v'(f(t)) = v''(f(t)) \in \Gamma. \tag{4.5.2.2}$$

*Proof*: We shall prove (4.5.2.2) by induction on the degree $n$ of $f(t)$. The case $n = 1$ is just (4.5.2.1). Suppose $n \geq 2$, and (4.5.2.2) is true for all degrees $< n$. Then we prove (4.5.2.2) for degree $n$. If $f$ is reducible in $F[t]$, then we obtain (4.5.2.2) from the induction hypothesis. So let $f$ be irreducible in $F[t]$, with $\deg f = n$. The valuations $v'$ and $v''$ of $F(t)$ agree with each other on the $n$-dimensional vector subspace (over $F$)

$$V = F + Ft + \cdots + Ft^{n-1}$$

of $F(t)$, by the induction hypothesis. Write $w$ for the (common) restriction $v'|_V = v''|_V$ of $v'$ and $v''$ to $V$. Then

$$w : V \to \Gamma \cup \{\infty\} \quad \text{is surjective,} \tag{4.5.2.3}$$

since $v'(F^\times)$ is already equal to $\Gamma$.

We define a "multiplication modulo $f$" on $V$, as usual. Then $V$ becomes isomorphic to the field

$$F_1 = F[t]/(f);$$

more precisely, every element of $F_1$ is of the form $g + (f)$, for a uniquely determined $g \in V$; we identify $g$ with $g + (f)$. Then (4.5.2.3) can be regarded as a mapping

$$w : F_1 \to \Gamma \cup \{\infty\}$$

which, if it satisfied the multiplicative property

$$w(g \cdot h) = w(g) + w(h)$$

for $g, h \in V$, would be an immediate extension of the valuation $v$ on $F$ to the algebraic extension field $F_1$. But this is not possible, by the hypothesis. Therefore there exist $g, h \in V$ with

$$w(r) \neq w(g) + w(h) = v'(gh), \tag{4.5.2.4}$$

where $r \in F[t]$ is the remainder of the polynomial $gh$ upon division by $f$. Using $\deg r < n = \deg f$, we get $r \in V$. The unique quotient $q \in F[t]$ satisfying the relation

$$gh = fq + r$$

has degree $\leq n-2 < n$, and thus also lies in $V$. From this and (4.5.2.4) we obtain

$$\begin{aligned} v'(f) &= v'(gh-r) - w(q) \\ &= \min\{v'(gh),\ w(r)\} - w(q) \\ &= \min\{w(g)+w(h),\ w(r)\} - w(q). \end{aligned} \tag{4.5.2.5}$$

Since, as already noted, $v'$ and $v''$ both agree with $w$ on $V$, we conclude that $v''(f)$ is also equal to (4.5.2.5), and hence equal to $v'(f)$, proving (4.5.2.2) for $f$. In particular, therefore,

$$v'(f) = v''(f). \qquad \square$$

*Remark 4.5.3.* Examples of valued fields $(F, \mathcal{O})$ satisfying the hypotheses of Lemma 4.5.2 include:

(1) $F$ algebraically closed and $\mathcal{O}$ arbitrary.
(2) $F$ real closed and $\mathcal{O}$ convex.
(3) $(F, \mathcal{O})$ Henselian and $\operatorname{char} \overline{F} = 0$.
(4) $(F, \mathcal{O})$ $p$-adically closed (cf. Exercise 4.7.1).

*Proof*: That example (1) satisfies the hypotheses is trivial. That examples (3) and (4) satisfy the hypotheses follows from Theorem 4.3.5. We immediately see that example (2) satisfies the hypotheses, as follows: since $\mathcal{O}$ is convex, $\overline{F}$ is ordered. In the only proper, algebraic extension field of $F$, the algebraic closure $\widetilde{F}$ of $F$, however, $-1 = a^2$ holds for some $a \in \widetilde{F}$. For an immediate extension of $\mathcal{O}$, then, $\overline{-1} = \overline{a}^2$ must hold in the residue field; but this is identical with $\overline{F}$. □

The next corollary follows by analogy with Corollaries 4.4.3 and 4.4.4:

**Corollary 4.5.4.** *The theory of real closed fields with nontrivial convex valuation ring is model complete and complete. It is the model companion of the theory of ordered fields with convex valuation ring.* □

## 4.6 Henselian Fields

In this section we present the model theory of Henselian fields. At the end, we shall give an application to the field of number theory. First, however, we convince ourselves that the class of Henselian fields is axiomatizable in the language of fields with the additional unary predicate $V$ (for a valuation ring).

To the axioms $K_1$–$K_9$ (from Section 1.6) and $V_1$–$V_3$ (from Section 4.4) we add, for every degree $n \geq 2$, the axiom $H_n$. In $H_n$ we shall express condition (3) of Theorem 4.3.2 for all monic polynomials $f$ of degree $n$. If

$$f = X^n + x_{n-1}X^{n-1} + \cdots + x_0,$$

then we shall, in $H_n$, utilize the expressions $(\forall f \in V[X])\,\varphi$ and $\psi(f(y))$, which are abbreviations for the expressions

$$\forall x_0,\dots,x_{n-1}\left[\bigwedge_{i=0}^{n-1} V(x_i) \to \varphi\right]$$

and $\psi(y^n + x_{n-1}y^{n-1} + \cdots + x_0)$, respectively. Furthermore, $f'$ will stand for the formal derivative

$$nX^{n-1} + (n-1)x_{n-1}X^{n-2} + \cdots + x_1$$

of $f$. Finally, $V^\times(z)$ will be the formula

$$V(z) \wedge \exists y\,(yz \doteq 1 \wedge V(y)),$$

which asserts that $z$ is a unit in $V$. We now put:

$$\begin{aligned}\mathrm{H}_n:\quad (\forall f \in V[X])\,\forall y\,[\neg V^\times(f(y)) \wedge V^\times(f'(y)) \wedge V(y) \to\\ \exists z\,(V(z) \wedge f(z) \doteq 0 \wedge \neg V^\times(y-z))].\end{aligned}$$

In the following we shall exploit the fact that we can also speak about the residue field $\overline{F}$ and the value group $\Gamma$ in the language of valued fields $(F,\mathscr{O})$, as follows.

For the residue fields we introduce the following "translation" $\varphi_r$ of a formula $\varphi$ of the field language, which we define by recursion on its construction:

$$\begin{aligned}(t_1 \doteq t_2)_r &:= \neg V^\times(t_1 - t_2)\\ (\neg\varphi)_r &:= \neg\varphi_r\\ (\varphi \wedge \psi)_r &:= (\varphi_r \wedge \psi_r)\\ (\forall x\,\varphi)_r &:= \forall x\,(V(x) \to \varphi_r)\end{aligned}$$

If $\mathrm{Fr}(\varphi) \subseteq \{v_0,\dots,v_n\}$, then this holds also for $\varphi_r$, and if $a_0,\dots,a_n \in \mathscr{O}$, then, for the valued field $(F,\mathscr{O})$ we have:

$$\overline{F} \models \varphi\,[\overline{a_0},\dots,\overline{a_n}] \quad \text{iff} \quad (F,\mathscr{O}) \models \varphi_r\,[a_0,\dots,a_r].$$

Readers may easily convince themselves of this using induction on the recursive construction of formulae.

We define the "translation" $\varphi_g$ of a formula $\varphi$ of the language of ordered groups as follows (where we write the group operation *multiplicatively* in this exceptional case):

$$\begin{aligned}(t_1 \doteq t_2)_g &:= \exists x\,(V^\times(x) \wedge xt_1 \doteq t_2)\\ (t_1 < t_2)_g &:= \exists x\,(V(x) \wedge \neg V^\times(x) \wedge xt_1 \doteq t_2)\end{aligned}$$

(in both cases $x$ should not occur in either $t_1$ or $t_2$)

$$\begin{aligned}(\neg\varphi)_g &:= \neg\varphi_g\\ (\varphi\wedge\psi)_g &:= (\varphi_g\wedge\psi_g)\\ (\forall x\,\varphi)_g &:= \forall x\,(x\neq 0\rightarrow\varphi_g)\end{aligned}$$

If $\mathrm{Fr}(\varphi)\subseteq\{v_0,\ldots,v_n\}$, then this holds also for $\varphi_g$, and if $a_0,\ldots,a_n\in F^\times$, then we have:

$$\Gamma\models\varphi\,[a_0\mathscr{O}^\times,\ldots,a_n\mathscr{O}^\times]\quad\text{iff}\quad(F,\mathscr{O})\models\varphi_g\,[a_0,\ldots,a_r].$$

This, too, is easily proved by induction on the recursive construction of formulae.

From our reflections above, the following emerges: if $\Phi$ is a type of $\overline{F}$ with certain constants $\overline{a}$, then $\Phi_r=\{\varphi_r\mid\varphi\in\Phi\}$ is a type of $(F,\mathscr{O})$ with certain representatives $a$ of the residue classes $\overline{a}$; and analogously for $\Gamma$. We therefore obtain:

If the valued field $(F,\mathscr{O})$ is $\kappa$-saturated,
then so are its residue field $\overline{F}$ and its value group $\Gamma$. (4.6.0.1)

The following theorem (as well as the Embedding Lemma 4.6.2) contains the essential kernel of the theorem of Ax-Kochen and Ershov (Theorem 4.6.4). The proof of the lemma is analogous to the proofs of Theorems 4.4.2 and 4.5.1.

**Theorem 4.6.1** (Ax-Kochen, Ershov). *Let $(F,\mathscr{O})$ be a Henselian valued field with value group $\Gamma$ and residue field $\overline{F}$. We assume* $\operatorname{char}\overline{F}=0$. *Then if $(F_1,\mathscr{O}_1)\supseteq(F,\mathscr{O})$ is a valued field with value group $\Gamma_1$ and residue field $\overline{F_1}$, and if $\Gamma$ is existentially closed (as an ordered group) in $\Gamma_1$, and $\overline{F}$ is existentially closed (as a field) in $\overline{F_1}$, then also $(F,\mathscr{O})$ is existentially closed in $(F_1,\mathscr{O}_1)$.*

*Proof*: Let $(F_2,\mathscr{O}_2)$ be an elementary, $\kappa^+$-saturated extension of $(F,\mathscr{O})$, where $\kappa=\operatorname{card}(F_1)$. Since $(F,\mathscr{O})$ is Henselian, so is $(F_2,\mathscr{O}_2)$. Such an extension exists by Theorem 2.5.2. We shall show that $(F_1,\mathscr{O}_1)$ embeds in $(F_2,\mathscr{O}_2)$ over $(F,\mathscr{O})$. Then by Corollary 2.5.5(2), $(F,\mathscr{O})$ will be existentially closed in $(F_1,\mathscr{O}_1)$.

By (4.6.0.1), the residue field $\overline{F_2}$ and the value group $\Gamma_2$ of $(F_2,\mathscr{O}_2)$ are likewise $\kappa^+$-saturated. From Corollary 2.5.5(1)(a) combined with the hypothesis of the theorem, we conclude that the residue field $\overline{F_1}$ and the value group $\Gamma_1$ of $(F_1,\mathscr{O}_1)$ embed into $\overline{F_2}$ and $\Gamma_2$ over $\overline{F}$ and $\Gamma$, respectively. By means of these embeddings we assume from now on that $\overline{F_1}\subseteq\overline{F_2}$ and $\Gamma_1\subseteq\Gamma_2$. We further assume that $(F_1,\mathscr{O}_1)$ is likewise Henselian. This can be achieved immediately by passing to the Henselization, which does not alter the hypotheses on the residue field and the value group.

Now we have the following situation:

$$\begin{array}{ccccc}(F_1,\mathscr{O}_1) & & & & (F_2,\mathscr{O}_2)\\ & \diagdown & & \diagup & \\ & (F,\mathscr{O}) & \xrightarrow{\text{id}} & (F,\mathscr{O}), & \end{array}$$

where $(F_1,\mathscr{O}_1)$ and $(F_2,\mathscr{O}_2)$ are Henselian, $(F_2,\mathscr{O}_2)$ is $\operatorname{card}(F_1)^+$-saturated, and the following conditions hold:

(i) $(F,\mathscr{O})$ is Henselian,

(ii) $\overline{F}\subseteq\overline{F_1}\subseteq\overline{F_2}$, and

(iii) $\Gamma \subseteq \Gamma_1 \subseteq \Gamma_2$ and
$\Gamma$ is *pure* in $\Gamma_1$, i.e. if $n\gamma_1 \in \Gamma$ for $\gamma_1 \in \Gamma_1$ and $n \geq 1$, then already $\gamma_1 \in \Gamma$.

We shall show that this is sufficient for the embedding id of $(F, \mathcal{O})$ to extend to an embedding of $(F_1, \mathcal{O}_1)$ into $(F_2, \mathcal{O}_2)$. With the help of Zorn's lemma (and after identification with the image), we may assume that the embedding id of $(F, \mathcal{O})$ is maximal with (i)–(iii). Then we have to show $F = F_1$. So we assume $F \neq F_1$, and distinguish three cases, as in the proof of Theorem 4.4.2.

*Case 1*: $\overline{F} \neq \overline{F_1}$. Let, say, $x_1 \in \mathcal{O}_1$ with $\overline{x_1} \in \overline{F_1} \setminus \overline{F}$. We first assume that $\overline{x_1}$ is transcendental over $\overline{F}$. Then pick $x_2 \in \mathcal{O}_2$ with $\overline{x_2} = \overline{x_1}$. (Here we write $\overline{\phantom{x}}$ for the residue class both in $\mathcal{O}_1$ and in $\mathcal{O}_2$, so long as this does not lead to collisions.) By Theorem 4.3.1(2), $x_1$ and $x_2$ are transcendental over $F$. Therefore, the mapping $a \mapsto a$ for $a \in F$ and $x_1 \mapsto x_2$ defines a monomorphism $\sigma$ from $F(x_1)$ into $F_2$ over $F$, which respects the valuations $\mathcal{O}_1$ and $\mathcal{O}_2$ as in Case 1 of the proof of (4.4.2). Namely, for $a_0, \ldots, a_n \in F$ and $k = 1, 2$, we just get

$$v_k(a_n x_k^n + \cdots + a_0) = \min_{0 \leq i \leq n} v(a_i),$$

as in (4.4.2.1). From this we see, in particular, that for the value group $\Gamma'$ and the residue field $\overline{F'}$ of the restriction $\mathcal{O}_k'$ of $\mathcal{O}_k$ to $F_k' = F(x_k)$, we have:

$$\Gamma' = \Gamma \quad \text{and} \quad \overline{F'} = \overline{F}(\overline{x_1}) = \overline{F}(\overline{x_2}).$$

We must still extend $\sigma$ to an algebraic extension of $F_1'$, in order to obtain (i) again.

For this we pass from $(F_k', \mathcal{O}_k')$ to the corresponding Henselization, which, by Theorem 4.3.3, we can assume to be a subfield of $(F_k, \mathcal{O}_k)$. By the uniqueness of the Henselization, $\sigma$ extends to it. We may therefore assume that (i) is still fulfilled for $(F_k', \mathcal{O}_k')$.

It remains to consider algebraic extensions of $\overline{F'}$ in $\overline{F_1}$. For this, let $\overline{\alpha} \in \overline{F_1} \setminus \overline{F}$ be algebraic over $\overline{F'}$. Let a monic polynomial $f_1(X) \in \mathcal{O}_1[X]$ be chosen so that $\overline{f_1} = \mathrm{Irr}\,(\overline{\alpha}, \overline{F'})$. Since $\mathrm{char}\,\overline{F_1} = 0$, $\overline{\alpha}$ is a simple zero of $\overline{f_1}$ in $\overline{F_1}$. Since $(F_1, \mathcal{O}_1)$ is Henselian, there is a $\beta_1 \in \mathcal{O}_1$ with $f_1(\beta_1) = 0$ and $\overline{\beta_1} = \overline{\alpha}$. Then the restriction of $\mathcal{O}_1$ to $F_1'(\beta_1)$ yields, using Theorem 4.3.1(2), the residue field $\overline{F_1'}(\overline{\alpha})$ and the value group $\Gamma$. If $f_2 = \sigma(f_1) \in \mathcal{O}_2[X]$, then, analogously, $\overline{\alpha}$ is a simple zero of $\overline{f_2}$ in $\overline{F_2}$. As before, there exists a $\beta_2 \in \mathcal{O}_2$ with $f_2(\beta_2) = 0$ and $\overline{\beta_2} = \overline{\alpha}$. Since both $f_1$ and $f_2$ are irreducible over $F_1'$ and $F_2'$, respectively, we obtain an extension

$$\tilde{\sigma} : F_1'(\beta_1) \to F_2'(\beta_2)$$

of $\sigma$, which, by the uniqueness of the extension of the valuation $\mathcal{O}_k'$ to $F_k'(x_k)$ (for $k = 1, 2$) is also valuation preserving.

We have thus, under the assumption that $\overline{F} \neq \overline{F_1}$, reached a contradiction to the maximality of the embeddability of $(F, \mathcal{O})$ into $(F_2, \mathcal{O}_2)$. Therefore $\overline{F} = \overline{F_1}$.

*Case 2*: $\overline{F} = \overline{F_1}$ and $\Gamma \neq \Gamma_1$. Let $\gamma \in \Gamma_1 \setminus \Gamma$ and $x \in F_1$ with $v_1(x) = \gamma$. By (iii), the sum $\Gamma + \mathbb{Z}\gamma$ is direct. Then for $a_0, \ldots, a_n \in F$ it follows, as in (4.4.2.2), that

$$v'(a_nx^n+\cdots+a_0)=\min_{0\le i\le n}(v(a_i)+i\gamma). \qquad (4.6.1.1)$$

Therefore the restriction $\mathcal{O}_1'$ of $\mathcal{O}_1$ to the intermediate field $F_1'=F(x)$ has $\Gamma\oplus\mathbb{Z}\gamma$ as value group; its residue field obviously remains $\overline{F_1}$. By Theorem 4.3.1(2), $x$ is transcendental over $F$.

Now, before we seek an embedding of $(F_1',\mathcal{O}_1')$ into $(F_2,\mathcal{O}_2)$ over $(F,\mathcal{O})$, we want to extend $F_1'$ algebraically far enough so that its value group is once again pure in $\Gamma_1$. We do this as follows. Let $\Gamma_1'$ be the value group of $(F_1',\mathcal{O}_1')$, and suppose $\Gamma_1'$ is not pure in $\Gamma_1$. Then there is a prime number $q$ and a $\delta\in\Gamma_1\setminus\Gamma_1'$ such that $q\delta\in\Gamma_1'$. Choose $y\in F_1$ with $v_1(y)=\delta$ and $a\in F_1'$ with $v_1'(a)=q\delta$. Then $v_1(y^qa^{-1})=0$. Therefore $y^qa^{-1}$ is a unit in $\mathcal{O}_1$, and, using $\overline{F_1}=\overline{F_1'}$, there is a $c\in F_1'$, $c\neq 0$, with $\overline{y^qa^{-1}}=\bar{c}$, i.e. $\overline{y^qa^{-1}c^{-1}}=\bar{1}$. Therefore, since $\operatorname{char}\overline{F_1}=0$, (the residue of) the polynomial $X^q-y^qa^{-1}c^{-1}\in\mathcal{O}_1[X]$ has a simple zero in $\overline{F_1}$. Then the Henselian property of $(F_1,\mathcal{O}_1)$ guarantees the existence of a $z\in\mathcal{O}_1^\times$ with $z^q=y^qa^{-1}c^{-1}$. Hence $ac$ is a $q$th power in $F_1$. It is therefore clear that $F_1'(yz^{-1})$ is an algebraic extension of $F_1'$, whose value group contains $v_1(yz^{-1})=v_1(y)-v_1(z)=\delta-0$. If we iterate this process sufficiently often (or simply apply Zorn's lemma), we finally obtain an algebraic extension $(F_1'',\mathcal{O}_1'')$ of $(F_1',\mathcal{O}_1')$ with $\mathcal{O}_1''=\mathcal{O}_1\cap F_1''$, whose value group $\Gamma_1''$ is pure in $\Gamma_1$. It remains to find a valuation-preserving embedding of the valued field $(F_2'',\mathcal{O}_1'')$ into $(F_1,\mathcal{O}_2)$ over $(F,\mathcal{O})$.

By the $\operatorname{card}(F_1)^+$-saturation of $(F_2,\mathcal{O}_2)$, it suffices, by Corollary 2.5.5(1)(b), to embed every finitely generated extension $F_1^*$ of $F$ in $F_1''$ with the induced valuation $\mathcal{O}_1^*$ into $(F_2,\mathcal{O}_2)$ over $(F,\mathcal{O})$. So let $F_1^*=F(x,y_1)$, where $y_1\in F_1''$ is algebraic over $F(x)$. Recall that the value group of $F(x)$ is just

$$\Gamma\oplus\mathbb{Z}v_1(x).$$

By Theorem 4.3.1(2), the value group $\Gamma_1^*$ of $F_1^*$ is a finite extension of $\Gamma\oplus\mathbb{Z}v_1(x)$. Then by the purity of $\Gamma$ in $\Gamma_1$, $\Gamma_1^*$ itself must have the form $\Gamma\oplus\mathbb{Z}\gamma_1$ for some $\gamma_1\in\Gamma_1$. If $x_1\in F_1^*$ is such that $v_1(x_1)=\gamma_1$, then $(F_1^*,\mathcal{O}_1^*)$ is obviously an immediate extension of $(F(x_1),\ \mathcal{O}_1\cap F(x_1))$. Namely, the valuation on $F(x_1)$ is, as in (4.6.1.1), given by

$$v_1(a_nx_1^n+\cdots+a_0)=\min_{0\le i\le n}(v(a_i)+i\gamma_1)$$

for $a_0,\dots,a_n\in F$; therefore the value group is $\Gamma\oplus\mathbb{Z}\gamma_1$.

Now it suffices to choose $x_2\in F_2$ such that $v_2(x_2)=\gamma_1$. It then again follows that

$$v_2(a_nx_2^n+\cdots+a_0)=\min_{0\le i\le n}(v(a_i)+i\gamma_1)$$

for $a_0,\dots,a_n\in F$. Therefore the mapping $a\mapsto a$ for $a\in F$ and $x_1\mapsto x_2$ defines a valuation-preserving embedding of $F(x_1)$ into $F_2$. As in Case 1, this embedding extends to the Henselization of $F(x_1)$ (which is contained in $F_1$). However, using $\operatorname{char}\overline{F_1}=0$ and Theorem 4.3.5, the Henselization is a maximal, immediate, algebraic extension of $(F(x_1),\ \mathcal{O}_1\cap F(x_1))$. We may therefore assume that $F_1^*$ is contained in the Henselization, and so obtain a valuation-preserving embedding of $F_1^*$

into $F_2$. Finally, we therefore also know that $F_1''$ has a valuation-preserving embedding into $F_2$ over $F$.

Properties (ii) and (iii) hold for $(F_1'', \mathcal{O}_1'')$. Property (i) can be achieved as in Case 1, by passing to the Henselization. In summary the assumption $\Gamma \neq \Gamma_1$ has led to a contradiction of the maximality of $F$. Therefore we must have $\Gamma = \Gamma_1$, after all.

*Case 3*: $\overline{F} = \overline{F_1}$ and $\Gamma = \Gamma_1$. Assume $F \neq F_1$. Let, say, $x_1 \in F_1 \setminus F$. Since $(F, \mathcal{O})$ is Henselian and $\operatorname{char}\overline{F_1} = 0$, the element $x_1$ must be transcendental over $F$, by Theorem 4.3.5. As in (4.4.2.3), we find an $x_2 \in F_2 \setminus F$ with

$$v_1(x_1 - a) = v_2(x_2 - a) \quad \text{for all } a \in F.$$

To conclude that $x_2$ is transcendental over $F$, we extend the type $\Phi(v_0)$ in (4.4.2.4) to

$$\Phi_1(v_0) = \Phi(v_0) \cup \{ f(x_0) \neq 0 \mid f \in F[X] \text{ irreducible and } \deg f > 1 \}.$$

Using Theorem 4.3.5 and Lemma 4.5.2, the mapping $a \mapsto a$ for $a \in F$ and $x_1 \mapsto x_2$ defines a valuation-preserving monomorphism of $F(x_1)$ into $F_2$, whose canonical extension to the Henselization again has properties (i)–(iii). This contradicts our maximality assumption. Finally, therefore, $F = F_1$. □

Looking ahead to further applications, we would like to isolate the embedding part of this proof. Strictly speaking, we have proved the following embedding lemma:

**Lemma 4.6.2** (Embedding lemma). *For $k = 1, 2$, let $(F_k, \mathcal{O}_k)$ be a Henselian field, with Henselian subfield $(F_k', \mathcal{O}_k')$. Further, let $\sigma' : (F_1', \mathcal{O}_1') \to (F_2', \mathcal{O}_2')$ be an isomorphism, and let $\sigma_r' : \overline{F_1'} \to \overline{F_2'}$ and $\sigma_g' : \Gamma_1' \to \Gamma_2'$ be the isomorphisms induced by $\sigma'$ on the residue field and the value group, respectively. Suppose $\Gamma_1'$ is pure in $\Gamma_1$. Then if $(F_2, \mathcal{O}_2)$ is $\operatorname{card}(F_1)^+$-saturated, and if $\sigma_r'$ and $\sigma_g'$ extend to embeddings $\sigma_r$ and $\sigma_g$ of $\overline{F_1}$ into $\overline{F_2}$ and of $\Gamma_1$ into $\Gamma_2$, respectively, then also $\sigma'$ extends to an embedding $\sigma$ of $(F_1, \mathcal{O}_1)$ into $(F_2, \mathcal{O}_2)$ that induces $\sigma_r$ and $\sigma_g$. Here it is assumed that $\operatorname{char}(\overline{F_1}) = 0$.* □

Using Theorem 4.6.1 and Robinson's Test 3.3.3, we obtain:

**Corollary 4.6.3.** *Let $T_r$ be a model complete theory of fields with characteristic 0, and $T_g$ a model complete theory of ordered Abelian groups. Then the theory $T$ of Henselian valued fields whose residue field is a model of $T_r$ and whose value group is a model of $T_g$, is itself model complete.* □

Recall the translations $\varphi \mapsto \varphi_r$ and $\varphi \mapsto \varphi_g$ that we defined at the beginning of this section. If $\Sigma_r$ is an axiom system for $T_r$, and $\Sigma_g$ an axiom system for $T_g$, then the theory $T$ described in the corollary can be axiomatized by:

(1) $K_1$–$K_9$,

(2) $V_1$–$V_3$, $H_n$ for $n \geq 2$,

(3) $\alpha_r$ for $\alpha \in \Sigma_r$,

(4) $\beta_g$ for $\beta \in \Sigma_g$.

The next theorem was proved independently by Ax-Kochen and Ershov. It is the main theorem in the model theory of Henselian fields.

**Theorem 4.6.4** (Ax-Kochen, Ershov). *Let* $(F_1, \mathcal{O}_1)$ *and* $(F_2, \mathcal{O}_2)$ *be Henselian fields with elementarily equivalent residue fields* $\overline{F_1}, \overline{F_2}$ *and elementarily equivalent value groups* $\Gamma_1, \Gamma_2$, *respectively. Then if the residue characteristic is* 0, *then also* $(F_1, \mathcal{O}_1)$ *and* $(F_2, \mathcal{O}_2)$ *are elementarily equivalent.*

*Proof*: The strategy of the proof is the following. First we assume, by applying Theorem 2.5.2, that $(F_k, \mathcal{O}_k)$ is $\aleph_1$-saturated ($k = 1, 2$). Then it is clear that in the passage to elementary extensions (or elementary substructures), the elementary equivalence of the residue fields and of the value groups will be preserved. Next, we construct, using Corollary 2.3.4, an ascending chain

$$\left(F_k^{(0)}, \mathcal{O}_k^{(0)}\right) \subseteq \cdots \subseteq \left(F_k^{(n)}, \mathcal{O}_k^{(n)}\right) \subseteq \cdots \subseteq (F_k, \mathcal{O}_k)$$

of countable Henselian fields $\left(F_k^{(n)}, \mathcal{O}_k^{(n)}\right)$, for $k = 1, 2$ and $n \in \mathbb{N}$, with isomorphisms

$$\sigma^{(n)} : \left(F_1^{(n)}, \mathcal{O}_1^{(n)}\right) \to \left(F_2^{(n)}, \mathcal{O}_2^{(n)}\right),$$

such that $\sigma^{(n+1)}$ extends $\sigma^{(n)}$. Moreover, the construction will be such that, for $n \geq 1$:

$$\left(F_1^{(2n-1)}, \mathcal{O}_1^{(2n-1)}\right) \preceq (F_1, \mathcal{O}_1), \tag{4.6.4.1}$$

$$\left(F_2^{(2n)}, \mathcal{O}_2^{(2n)}\right) \preceq (F_2, \mathcal{O}_2), \tag{4.6.4.2}$$

$$\left(\overline{F_1}, (\overline{a})_{a \in \mathcal{O}_1^{(n)}}\right) \equiv \left(\overline{F_2}, \left(\overline{\sigma^{(n)}(a)}\right)_{a \in \mathcal{O}_1^{(n)}}\right), \tag{4.6.4.3}$$

$$\left(\Gamma_1, (v_1(a))_{a \in F_1^{(n)}}\right) \equiv \left(\Gamma_2, (v_2(\sigma^{(n)}(a)))_{a \in F_1^{(n)}}\right). \tag{4.6.4.4}$$

The conditions (4.6.4.3) and (4.6.4.4) will enable the induction to proceed.

If we set

$$F_1' = \bigcup_n F_1^{(n)}, \quad F_2' = \bigcup_n F_2^{(n)}, \quad \sigma' = \bigcup_n \sigma^{(n)},$$

then we obtain the following situation:

$$\begin{array}{ccc} (F_1, \mathcal{O}_1) & & (F_2, \mathcal{O}_2) \\ \rotatebox{90}{$\preceq$} & & \rotatebox{90}{$\preceq$} \\ (F_1', \mathcal{O}_1') & \xrightarrow{\sigma'} & (F_2', \mathcal{O}_2'). \end{array}$$

Here, the elementary inclusions

$$(F_k', \mathcal{O}_k') \preceq (F_k, \mathcal{O}_k) \qquad (k = 1,2),$$

follow from

$$\bigcup_n F_k^{(2n)} = \bigcup_n F_k^{(2n-1)} = F_k' \qquad (k = 1,2),$$

(4.6.4.1), (4.6.4.2), and Theorem 2.4.6. Hence we obtain, finally, the desired elementary equivalence of $(F_1, \mathcal{O}_1)$ with $(F_2, \mathcal{O}_2)$.

For $n = 0$ let $F_1^{(0)} = F_2^{(0)} = \mathbb{Q}$ and $\sigma^{(0)} = \mathrm{id}_{\mathbb{Q}}$. Since $\operatorname{char} \overline{F} = 0$, we then also have $\mathcal{O}_1^{(0)} = \mathcal{O}_2^{(0)} = \mathbb{Q}$; in particular, $\left(F_k^{(0)}, \mathcal{O}_k^{(0)}\right)$ is a Henselian field, for $k = 1,2$.

For $n = 1$, let $\left(F_1^{(1)}, \mathcal{O}_1^{(1)}\right)$ be a countable, elementary substructure of $(F_1, \mathcal{O}_1)$ containing $\left(F_1^{(0)}, \mathcal{O}_1^{(0)}\right)$. Such a substructure exists, by Theorem 2.3.3. The countability of $F_1^{(1)}$ implies that of its residue field $\overline{F_1^{(1)}}$ and value group $\Gamma_1^{(1)}$. Moreover, we have:

$$\overline{F_1^{(1)}} \equiv \overline{F_1} \equiv \overline{F_2} \quad \text{and} \quad \Gamma_1^{(1)} \equiv \Gamma_1 \equiv \Gamma_2.$$

Therefore, by the $\aleph_1$-saturation of $\overline{F_2}$ and $\Gamma_2$, there are elementary embeddings

$$\sigma_r^{(1)} : \overline{F_1^{(1)}} \to \overline{F_2} \quad \text{and} \quad \sigma_g^{(1)} : \Gamma_1^{(1)} \to \Gamma_2,$$

by Corollary 2.5.6. Since $\Gamma_1^{(0)} = \{0\}$ is pure in $\Gamma_1^{(1)}$, the identity map

$$\sigma^{(0)} : \left(F_1^{(0)}, \mathcal{O}_1^{(0)}\right) \to \left(F_2^{(0)}, \mathcal{O}_2^{(0)}\right)$$

extends to an isomorphism

$$\sigma^{(1)} : \left(F_1^{(1)}, \mathcal{O}_1^{(1)}\right) \to \left(F_2^{(1)}, \mathcal{O}_2^{(1)}\right) \subseteq (F_2, \mathcal{O}_2)$$

such that $\sigma_r^{(1)}(\overline{a}) = \overline{\sigma^{(1)}(a)}$ and $\sigma_g^{(1)}(v_1(a)) = v_2\left(\sigma^{(1)}(a)\right)$ for $a \in F_1^{(1)}$, by the Embedding Lemma 4.6.2. Hence, in particular, $\sigma^{(1)}$ satisfies conditions (4.6.4.3) and (4.6.4.4).

The transition from $n = 1$ to $n = 2$ already illustrates the general transition from $n$ to $n+1$, where, for even $n$, the isomorphism $\sigma^{(n+1)}$ is to be defined, whilst for odd $n$, the isomorphism $\left(\sigma^{(n+1)}\right)^{-1}$ is to be defined. In our concrete case, therefore, we must define $\left(\sigma^{(2)}\right)^{-1}$.

In the case $n = 1$ we defined $\sigma^{(1)}$ such that (4.6.4.3) and (4.6.4.4) held, i.e. in particular:

$$\left(\overline{F_1}, (\alpha)_{\alpha \in \overline{F_1^{(1)}}}\right) \equiv \left(\overline{F_2}, \left(\sigma_r^{(1)}(\alpha)\right)_{\alpha \in \overline{F_1^{(1)}}}\right), \quad \text{and}$$

$$\left(\Gamma_1, (\gamma)_{\gamma \in \Gamma_1^{(1)}}\right) \equiv \left(\Gamma_2, \left(\sigma_g^{(1)}(\gamma)\right)_{\gamma \in \Gamma_1^{(1)}}\right).$$

With the help of Theorem 2.3.3 we choose $(F_2^{(2)}, \mathcal{O}_2^{(2)})$ as a countable substructure of $(F_2, \mathcal{O}_2)$ containing $(F_2^{(1)}, \mathcal{O}_2^{(1)})$. Then, in particular:

$$\left(\overline{F_1}, (\alpha)_{\alpha \in \overline{F_1^{(1)}}}\right) \equiv \left(\overline{F_2^{(2)}}, \left(\sigma_r^{(1)}(\alpha)\right)_{\alpha \in \overline{F_1^{(1)}}}\right), \quad \text{and}$$

$$\left(\Gamma_1, (\gamma)_{\gamma \in \Gamma_1^{(1)}}\right) \equiv \left(\Gamma_2^{(2)}, \left(\sigma_g^{(1)}(\gamma)\right)_{\gamma \in \Gamma_1^{(1)}}\right).$$

By Corollary 2.5.6 there are elementary embeddings

$$\left(\sigma_r^{(2)}\right)^{-1} : \overline{F_2^{(2)}} \to \overline{F_1} \quad \text{and} \quad \left(\sigma_g^{(2)}\right)^{-1} : \Gamma_2^{(2)} \to \Gamma_1$$

extending $(\sigma_r^{(1)})^{-1}$ and $(\sigma_g^{(1)})^{-1}$, respectively. Finally, since $\Gamma_2^{(1)} = \sigma_g^{(1)}(\Gamma_1^{(1)})$, as an elementary substructure of $\Gamma_2$ and thus also of $\Gamma_2^{(2)}$, is pure in $\Gamma_2^{(2)}$, we can apply the Embedding Lemma 4.6.2 in order to obtain an extension

$$\left(\sigma^{(2)}\right)^{-1} : \left(F_2^{(2)}, \mathcal{O}_2^{(2)}\right) \to \left(F_1^{(2)}, \mathcal{O}_1^{(2)}\right) \subseteq (F_1, \mathcal{O}_1)$$

of $(\sigma^{(1)})^{-1}$ that induces $(\sigma_r^{(2)})^{-1}$ and $(\sigma_g^{(2)})^{-1}$. □

Finally, we would like to give an application of Theorem 4.6.4 to the so-called "Artin Conjecture." This application contributed especially to the success of the model theory of Henselian fields.

A field $F$ is called a *$C_i$-field* if for every $d \geq 1$, every homogeneous polynomial of degree $d$ in more than $d^i$ variables over $F$ represents zero nontrivially over $F$ (i.e. has a nontrivial zero in $F$). It is known that all finite fields $\mathbb{F}_q$ are $C_1$-fields, and the field $\mathbb{F}_q((t))$ of formal Laurent series over any finite field $\mathbb{F}_q$ is a $C_2$-field. *Artin's conjecture* was that every $p$-adic number field $\mathbb{Q}_p$ is a $C_2$-field. This conjecture is suggested by the relative similarity of $\mathbb{F}_q((t))$ and $\mathbb{Q}_p$. Both fields are Henselian valued, and in both cases $\mathbb{Z}$ is the value group and $\mathbb{F}_p$ is the residue field. The essential distinction between the two fields lies in the characteristic: while the characteristic of $\mathbb{F}_q((t))$ is prime (namely, $p$), $\mathbb{Q}_p$ has characteristic 0. This distinction is removed when we take ultraproducts. The next theorem is based on this and on Theorem 4.6.4.

**Theorem 4.6.5** (Ax-Kochen: Affirmation of Artin's Conjecture for fixed degree and almost all $p$). *For every degree $d \geq 1$ there exists a lower bound $n_d$ such that for $p \geq n_d$, every homogeneous polynomial of degree $d$ in more than $d^2$ variables over $\mathbb{Q}_p$ represents zero nontrivially in $\mathbb{Q}_p$.*

*Proof*: Suppose that for some $d$ there were no such bound. Then there would exist an infinite subset $B$ of the set $\mathbb{P}$ of all prime numbers such that for every $p \in B$ there is a homogeneous polynomial $f_p$ of degree $d$ in more than $d^2$ variables over $\mathbb{Q}_p$ that has only the trivial zero in $\mathbb{Q}_p$. One easily sees that by suitable substitutions we can arrange that all polynomials $f_p$ begin with $X_1^p$. Now if we set sufficiently many variables $X_i$ (with $i \neq 1$) equal to 0, then we can assume that $f_p$ is a polynomial in the variables $X_1, \ldots, X_{d^2+1}$. If we write

$$f(Y_1, \ldots, Y_m; X_1, \ldots, X_{d^2+1}) \in \mathbb{Z}[Y_1, \ldots, Y_m; X_1, \ldots, X_{d^2+1}]$$

for the general homogeneous polynomial of degree $d$ in the variables $X_1, \ldots, X_{d^2+1}$ with indeterminate coefficients $Y_1, \ldots, Y_m$ (where $m = \binom{d^2+d}{d}$), then for all $p \in B$ the following elementary sentence $\alpha$ in the language of fields holds in $\mathbb{Q}_p$:

$$\exists y_1, \ldots, y_m \left[ \left( \bigvee_{i=1}^{m} y_i \neq 0 \right) \wedge \forall x_1, \ldots, x_n \left( f(y_1, \ldots, y_m; x_1, \ldots, x_n) \doteq 0 \rightarrow \bigwedge_{j=1}^{n} x_j \doteq 0 \right) \right],$$

where $n = d^2 + 1$.

Let $\mathscr{D}_0$ be the filter of all cofinite subsets of $\mathbb{P}$. Then obviously

$$\mathscr{D}_0 \cup \{ U \cap B \mid U \in \mathscr{D}_0 \}$$

is a system of nonempty subsets of $\mathbb{P}$ that is closed under finite intersection. By Lemma 2.6.1, it can be extended to an ultrafilter $\mathscr{D}$ on $\mathbb{P}$. Since $B \in \mathscr{D}$, Theorem 2.6.2 implies that the sentence $\alpha$ holds also in the ultraproduct

$$F_1 = \prod_{p \in \mathbb{P}} \mathbb{Q}_p \Big/ \mathscr{D}.$$

Since, as we mentioned above, all fields $\mathbb{F}_p((t))$ are $C_2$-fields, $\neg\alpha$ holds in $\mathbb{F}_p((t))$ for all $p \in \mathbb{P}$. Consequently, $\neg\alpha$ holds in the ultraproduct

$$F_2 = \prod_{p \in \mathbb{P}} \mathbb{F}_p((t)) \Big/ \mathscr{D}.$$

But, as we shall soon see, $F_1$ and $F_2$ are elementarily equivalent. Hence we have arrived at a contradiction to the supposition of the nonexistence of a bound $n_d$.

In order to see $F_1 \equiv F_2$, we form the ultraproducts of the fields considered above together with their canonical valuation rings; i.e.

$$(F_1, \mathscr{O}_1) = \prod_{p \in \mathbb{P}} (\mathbb{Q}_p, \mathbb{Z}_p) \Big/ \mathscr{D} \qquad \text{and}$$

$$(F_2, \mathscr{O}_2) = \prod_{p \in \mathbb{P}} (\mathbb{F}_p((t)), \mathbb{F}_p[[t]]) \Big/ \mathscr{D}.$$

The value group of these two valued fields is, in each case, the ultraproduct

$$\prod_{p\in\mathbb{P}} \mathbb{Z}/\mathscr{D} = \mathbb{Z}^{\mathbb{P}}/\mathscr{D}.$$

The residue field is in both cases the ultraproduct

$$\prod_{p\in\mathbb{P}} \mathbb{F}_p/\mathscr{D}.$$

Since the "characteristic sentence"

$$\mathsf{C}_p: \quad \underbrace{1+\cdots+1}_{p \text{ times}} \doteq 0$$

(1.6.1.4) holds only in the factor $\mathbb{F}_p$, the characteristic of this residue field is zero, by Theorem 2.6.2. Therefore, Theorem 4.6.4 implies the elementary equivalence of $(F_1, \mathscr{O}_1)$ and $(F_2, \mathscr{O}_2)$, whence, in particular, that of $F_1$ and $F_2$. □

With this theorem, Ax and Kochen proved in 1965 that for fixed degree, Artin's conjecture holds in almost all $\mathbb{Q}_p$. For the case of quadratic forms, i.e. $d = 2$, it had long been known that this holds even in *all* $\mathbb{Q}_p$. The corresponding result for $d = 3$ was proved in 1952 by D.J. Lewis. For $d \geq 4$, however, Theorem 4.6.5 proved to be the best possible result. Namely, in 1966 Terjanian proved that the form

$$f(X_1,X_2,X_3) + f(Y_1,Y_2,Y_3) + f(Z_1,Z_2,Z_3) + \\ 4f(U_1,U_2,U_3) + 4f(V_1,V_2,V_3) + 4f(W_1,W_2,W_3),$$

with

$$\begin{aligned} f(X_1,X_2,X_3) = {} & X_1^4 + X_2^4 + X_3^4 \\ & - X_1^2X_2^2 - X_1^2X_3^2 - X_2^2X_3^2 \\ & - X_1^2X_2X_3 - X_1X_2^2X_3 - X_1X_2X_3^2, \end{aligned}$$

has no nontrivial zero in $\mathbb{Q}_2$. Thus, for degree 4, Artin's conjecture does not hold in all $\mathbb{Q}_p$ (specifically, $\mathbb{Q}_2$ is not a $C_2$-field).

## 4.7 Exercises for Chapter 4

**4.7.1.** A valued field $(F, \mathscr{O})$ is called *p-adically closed* (cf. [Prestel–Roquette, 1984]; there designated "with $p$-rank 1") if the following hold:

(i) $\operatorname{char} F = 0$ and $(F, \mathscr{O})$ is Henselian,
(ii) the residue field $\overline{F}$ is the field $\mathbb{F}_p$ with $p$ elements, and
(iii) the value group $\Gamma$ is a $\mathbb{Z}$-group in which the value of $p$ is minimal positive.

The $p$-adic numbers $\mathbb{Q}_p$ with the $p$-adic valuation are $p$-adically closed.

Show that the class of $p$-adically closed fields is model complete and axiomatizable in the language of valued fields.

*Hints*: Apply Theorems 4.1.3 and 4.3.5, and show that the proof of Theorem 4.6.1 carries over to the case of two $p$-adically closed fields $(F,\mathscr{O}) \subseteq (F_1,\mathscr{O}_1)$. For this, observe that Case 1 of that proof does not arise, and that in Case 2 the hypothesis that $\operatorname{char}\overline{F} = 0$ is used only to show that every $b \in \mathscr{O}_1$ with $v_1(b-1) > 0$ is a $q$th power in $F_1$. This holds also in a $p$-adically closed field for every prime number $q \neq p$. For $q = p$, apply the fact that $b \in \mathscr{O}_1$ is a $p$th power provided $v_1(b-1) > 2v_1(p)$.

**4.7.2.** Show that the relative algebraic closure of a subfield $F$ in a $p$-adically closed field $(F_1,\mathscr{O}_1)$, with the induced valuation, is itself again $p$-adically closed. Conclude, using Exercise 4.7.1, that the theory of $p$-adically closed fields is complete, and hence equal to that of $\mathbb{Q}_p$.

**4.7.3.** The following theorem is known (cf. [Prestel–Roquette, 1984, 3.11]): If $(F_1,\mathscr{O}_1)$ and $(F_2,\mathscr{O}_2)$ are $p$-adically closed fields with the common valued subfield $(F,\mathscr{O})$, and if $a \in F$ is an $n$th power in $F_1$ if and only if it is an $n$th power in $F_2$ ($n \in \mathbb{N}$), then the relative algebraic closures of $F$ in $F_1$ and in $F_2$, with the corresponding, induced valuations, are isomorphic over $F$.

Show, with the help of this theorem and of Exercise 4.7.1, that the theory of $p$-adically closed fields in the language of fields with a valuation divisibility symbol together with unary predicate symbols $P_n$ ($n \geq 2$) and the additional axioms

$$\forall x\,(P_n(x) \leftrightarrow \exists y\; x \doteq y^n),$$

admits quantifier elimination.

# Appendix A
# Remarks on Decidability

In this appendix we would like to describe how one can make precise the concepts (used repeatedly in the Introduction) of "decidable" and "effectively enumerable". We would also like to show how a complete, effectively enumerable axiom system leads to a decidable theory. Finally, we would like to sketch a proof, going back to A. Tarski, of Gödel's first incompleteness theorem.

When making the concepts of "decidable" and "effectively enumerable" precise, one usually relies on the concept of "algorithm", which, in turn, can again be made precise with the help of discretely operating machines. As an idealization of such a discretely operating machine, one often utilizes the "Turing machine", which we now briefly describe. It has been proved that all idealizations of such machines considered so far, and, moreover, all proposed formalizations of the concept of "algorithm", are equivalent to each other. Considering this, the use of the concept of a Turing machine does not represent any restriction. At this point, we refer the interested reader to the corresponding literature – e.g. [Hermes, 1965–69].

We consider a language $L$ with finite index sets $I$, $J$ and $K$. We generate the variables $v_n$ ($n \in \mathbb{N}$) from the basic symbols $v$ and $'$ by identifying $v_n$ with the string of symbols

$$v\underbrace{\prime\prime\cdots\prime}_{n \text{ times}}.$$

We have thereby arranged for the alphabet $A$ of the language $L$ to be finite – say, $A = \{a_1, \ldots, a_n\}$. Later we shall also utilize the empty symbol, which we shall denote by $a_0$ (which, we assume, is not a member of $A$). Note that all the theories considered in Section 1.6 were axiomatized in languages that possess, in the sense described above, a finite alphabet.

We call a subset $M$ of the set $\mathrm{Sent}(L)$ of sentences *effectively enumerable* if there is a Turing machine that produces (or prints out) exactly the elements of $M$, in some sequence. We call $M$ *decidable* if there is a Turing machine which, upon receiving as input an arbitrary $L$-sentence $\alpha$, prints out (after finitely many steps of execution) the symbol $\doteq$ in case $\alpha$ lies in $M$, and the symbol $\neg$ in case $\alpha$ does not lie in $M$.

A. Prestel, C.N. Delzell, *Mathematical Logic and Model Theory*, Universitext, 
DOI 10.1007/978-1-4471-2176-3, 

Here, by a *Turing machine* on the alphabet $A = \{a_1, \ldots, a_n\}$, we mean a tape that is infinitely long in both directions, with discrete fields:

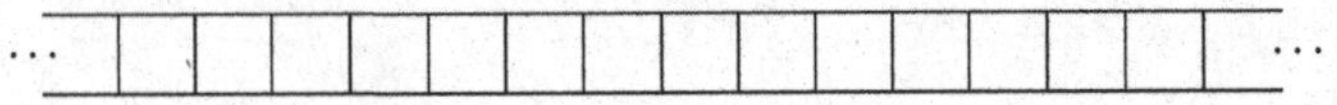

,

together with a "read–write head" and a program table. Each field of the tape either is empty (i.e. the empty symbol $a_0$ is there) or has been printed with exactly one symbol $a$ from $A$. At every step of execution, the read–write head can read the "working field" of this tape, and can perform any one of the following, finitely many possible steps of execution:

$$\begin{aligned} a_\nu &: \ \text{print } a_\nu \ (0 \leq \nu \leq n) \text{ on the working field,}^1 \\ r &: \ \text{move one field to the right,} \\ l &: \ \text{move one field to the left,} \\ s &: \ \text{stop.} \end{aligned}$$

Finally, the program table consists of a finite sequence of quadruples of the form

$$z \quad a \quad b \quad z'$$

where $a$ is an element of $A$, $b$ is a possible execution step of the work head, and $z, z'$ are elements of the finite set $Z = \{z_1, \ldots, z_m\}$. The elements of $Z$ are called "states"; they can, alternatively, be presented by the numerals 1 through $m$.

The manner of operation of our Turing machine is uniquely determined by its program table, as follows: if the machine is in state $z$, and if the input of the working field is $a$, then the work head reacts with $b$, and the machine goes into state $z'$. The finite, four-column matrix of which the program table consists, uniquely determines the manner of execution of the machine via the mapping

$$(z, a) \mapsto (b, z').$$

We may therefore identify the machine with its program table.

The axiom system of each of the theories given in Section 1.6 is, in the above precise sense, decidable. This is clear, since each such axiom system either is finite, or consists of axioms of a single schema; either case easily permits the construction of a Turing machine that tests whether any given sentence is an axiom or not. Hence these axiom systems are, in particular, effectively enumerable. Indeed, in general any decidable set of sentences is effectively enumerable: one builds a Turing machine that produces all sentences of the language under consideration in some sequence, but which prints out only those sentences that belong to $M$. Here, membership in $M$ is tested by coupling with a second Turing machine, whose existence is guaranteed by the assumed decidability of $M$.

The converse of the above observation does not hold: there are effectively enumerable sets that are not decidable (cf. [Hermes, 1965–69]).

[1] Overwriting any symbol that may have been in that field before.

Now, if $\Sigma$ is an effectively enumerable axiom system, consisting of $L$-sentences, then $\mathrm{Ded}(\Sigma)$ is likewise effectively enumerable. This is rather obvious, once one recalls the nature of the logical axioms and rules that we introduced in Section 1.3. One must construct a Turing machine that produces all deductions from $\Sigma$ in some sequence, where the axioms of $\Sigma$ are, in turn, furnished by a second Turing machine, whose existence is guaranteed by the hypothesis on $\Sigma$. Although this might seem not too difficult at first glance, carrying it out completely is rather laborious. Here we merely state this fact.

With this fact, however, the following theorem is easily proved:

**Theorem A.1.** *Let $\Sigma$ be an effectively enumerable, consistent axiom system of $L$-sentences. If $\Sigma$ is complete, then* $\mathrm{Ded}(\Sigma)$ *is a decidable theory.*

In the proof, we are given a Turing machine $T$ that effectively enumerates $\mathrm{Ded}(\Sigma)$. Now if $\alpha$ is an arbitrary, given $L$-sentence, then we know, by the completeness of $\Sigma$, that either $\alpha$ or $\neg\alpha$ belongs to $\mathrm{Ded}(\Sigma)$. We therefore need only construct a Turing machine $T'$ that compares, one at a time, the sentences produced by $T$ with $\alpha$ and with $\neg\alpha$. After finitely many steps, $T$ will print out either $\alpha$ or $\neg\alpha$. We program $T'$ so that in the first case, it prints out $\doteq$, and in the second case the symbol $\neg$. Thus the theorem is proved.

Theorem A.1 guarantees the decidability of a whole series of axiom systems whose completeness we have proven throughout this book. Recalling, in addition, that the completeness of $\Sigma$ implies

$$\mathrm{Th}(\mathfrak{A}) = \mathrm{Ded}(\Sigma)$$

for every model $\mathfrak{A}$ of $\Sigma$ (1.6.0.3), we deduce, among other things, the decidability of the theories of the following structures in the corresponding languages:

(1) the ordered, additive group of the integers $\mathbb{Z}$,
(2) the ordered field of real numbers $\mathbb{R}$,
(3) the field of complex numbers $\mathbb{C}$,
(4) the algebraic closure of the prime field $\mathbb{F}_p$ with $p$ elements, and
(5) the valued field of $p$-adic numbers $\mathbb{Q}_p$ (cf. Section 4.7).

Now we would like to close this appendix with a sketch of a proof of the undecidability of the theory of

$$\mathfrak{N} = \langle \mathrm{N}; +^{\mathrm{N}}, \cdot^{\mathrm{N}}; 0^{\mathrm{N}}, 1^{\mathrm{N}} \rangle.$$

Then, by Theorem A.1, there can be no effectively enumerable axiom system $\Sigma$ of sentences of the language of $\mathfrak{N}$ for which

$$\mathrm{Ded}(\Sigma) = \mathrm{Th}(\mathfrak{N}).$$

In particular, therefore, for the Peano axiom system $\Sigma_{\mathrm{PA}}$ given in Section 1.6, we have

$$\mathrm{Ded}(\Sigma) \subsetneq \mathrm{Th}(\mathfrak{N}).$$

We have thus obtained

**Theorem A.2** (Gödel's First Incompleteness Theorem). *There is a sentence $\alpha$ that is true in $\mathfrak{N}$, but is not deducible from $\Sigma_{\mathrm{PA}}$. Thus, the Peano axiom system is incomplete.* □

The proof of the undecidability of $\mathrm{Th}(\mathfrak{N})$ is indirect. We first assume that there *were* a Turing machine $T$ which, for any given sentence $\alpha$, decides whether $\mathfrak{N} \models \alpha$ or $\mathfrak{N} \not\models \alpha$ holds. This Turing machine is given by a finite matrix – its program table. Next, we encode this program table by a natural number, and simulate the behaviour of $T$ via number theoretic functions. For this, we may proceed as follows (say):

First we give a one-to-one correspondence between the elements of the disjoint union

$$A \,\dot{\cup}\, Z \,\dot{\cup}\, \{a_0, r, l, s\} \tag{A.2.1}$$

and (finitely many) natural numbers – their code numbers. Then we assign to every finite sequence of "symbols" belonging to the set (A.2.1) a natural number (its code) in such a way that the sequence of symbols can be uniquely and effectively reconstructed from this number. If we assume, e.g. that the symbols $\neg, v_0, \doteq, c_0$ are assigned the code numbers 2, 4, 1, 3, respectively, then:

$$\text{the formula } \neg v_0 \doteq c_0 \text{ can be encoded by } 2^2 \cdot 3^4 \cdot 5^1 \cdot 7^3 \ \ (= 555{,}660).$$

From the prime-power factorization of 555,660 we could read off the exponents of the prime numbers (in increasing order), and thereby reconstruct the formula $\neg v_0 \doteq c_0$. We denote the code number so assigned to a string $\zeta$ of symbols by $\ulcorner \zeta \urcorner$.

We can, now, in particular, assign in this way a natural number to every line of the program table of $T$. Repeating this process one more time, we can assign another natural number to the finite sequence of lines of this program table; from this number the entire program table could be reconstructed.

Finally, we show that the execution of a computation of the Turing machine $T$ can be described by an arithmetic function. Here we call a function $f(n) = m$ of natural numbers *arithmetic* if there is a formula $\varphi(v_0, v_1)$ of the language of $\mathfrak{N}$ such that, for $n, m \in \mathbb{N}$:

$$f(n) = m \quad \text{iff} \quad \mathfrak{N} \models \varphi(n, m).$$

In this case we call $\varphi$ a *defining formula for $f$*. Now if $f(n)$ is a function that assigns the number 1 (which is the code of $\doteq$) to the code $\ulcorner \rho \urcorner$ in case $\rho$ holds in $\mathfrak{N}$, and the number 2 (which is the code of $\neg$) in case $\rho$ does not hold in $\mathfrak{N}$, then there is a formula $\varphi$ of the language of $\mathfrak{N}$ with the property:

$$\mathfrak{N} \models \varphi(\ulcorner \rho \urcorner, 1) \quad \text{iff} \quad \mathfrak{N} \models \rho.$$

Thus we have the possibility to "speak" in $\mathfrak{N}$ about the truth of sentences (in $\mathfrak{N}$). This opens the possibility of simulating the liar's paradox. But to do this, we must make a small alteration.

It is easy to see that if $f$ is arithmetic, then so is the function $g$ defined as follows on the code $\ulcorner \rho \urcorner$ of a formula $\rho$ in the distinguished variable $v_0$:

$$g(\ulcorner\rho\urcorner) = \begin{cases} 1 & \text{if } \mathfrak{N} \models \neg\rho(\ulcorner\rho\urcorner) \\ 2 & \text{if } \mathfrak{N} \models \rho(\ulcorner\rho\urcorner). \end{cases}$$

Now if $\psi$ is a formula defining $g$, then for all formulae $\rho$ in the variable $v_0$, we have:

$$\mathfrak{N} \models \psi(\ulcorner\rho\urcorner, 1) \quad \text{iff} \quad \mathfrak{N} \models \neg\rho(\ulcorner\rho\urcorner). \tag{A.2.2}$$

Applying (A.2.2) to the formula

$$\rho_0(v_0) = \psi(v_0, 1),$$

we arrive at the contradiction

$$\mathfrak{N} \models \rho_0(\ulcorner\rho_0\urcorner) \quad \text{iff} \quad \mathfrak{N} \models \neg\rho_0(\ulcorner\rho_0\urcorner).$$

This contradiction arose, ultimately, from the assumption that $\mathrm{Th}(\mathfrak{N})$ is decidable. We have thus proved the undecidability of $\mathrm{Th}(\mathfrak{N})$, and consequently the incompleteness of the Peano axiom system.

In closing, we should mention that by similar methods one can prove the incompleteness of the Zermelo–Fraenkel axioms for set theory given in Section 1.6.

# Appendix B
# Remarks on Second-Order Logic

The attentive reader will have noticed that up to now, whenever we have spoken of "logic" in this text, we have always spoken only of first-order logic. The mention of a first order suggests the existence of a second order, and perhaps even higher orders. What, then, is the distinctive property of the first order? We shall discuss this question in this appendix.

One usually says that in the first order, one may quantify only over elements of a universe $A$, and not over subsets thereof or functions thereon. If we have two kinds of quantifiers – one for elements of $A$ and the other for subsets of $A$ –, then we usually say that we have a second-order logic (more precisely, second-order monadic logic, since here we quantify only over unary relations (i.e. subsets), and not over $n$-place relations for $n > 1$). Usually the first-order quantifiers $\forall x$ are written with lowercase variables, and second-order quantifiers (i.e. those that run over subsets of $A$) are written with uppercase letters: $\forall X$. However, this typographical distinction is obviously irrelevant. One could just as well write variables of the second kind using lowercase letters. Thus, there is no *syntactic* difference between quantifiers over elements and quantifiers over subsets. With no additional effort one could even introduce quantifiers over arbitrarily many sorts of objects, without thereby genuinely passing outside the first order. In second-order monadic logic, these sorts are, first, the elements of $A$, and, second, the subsets of $A$.

The actual content of the so-called "second order" lies in the semantics established on the above-mentioned syntax. Specifically, in second-order logic, there is an intimate, semantical connection between the first and second sorts; namely, one requires that the set-variables should range over "all" subsets of $A$; i.e. the second sort is the full power set of the first sort. As we shall see, this semantical connection between the two sorts can no longer be expressed in first-order logic: the properties of a power set cannot be expressed in first-order logic. The situation is similar to that of "well-orderedness": this, too, cannot be expressed in first-order logic; cf. Exercise 1.7.15.

We shall now make the above statements more explicit, in the following three steps:

A. Prestel, C.N. Delzell, *Mathematical Logic and Model Theory*, Universitext,
DOI 10.1007/978-1-4471-2176-3, 

*Step 1*: Here we shall introduce the syntax and semantics of second-order monadic logic. The set-variables will run over *all* subsets of the universe $A$ of individuals. In such a framework, the field of real numbers, for example, can be axiomatized up to isomorphism. From this it will follow that there can be no finitary, complete deductive system in this logic, since otherwise a Finiteness Theorem (analogous to (1.5.6)) would have to hold, which would imply that we would not be able to axiomatize any infinite structure up to isomorphism.

*Step 2*: Here we shall consider the two kinds of variables of second-order monadic logic, equally and independently of one another. That is, we introduce a first-order logic with two sorts of variables. For this there will be a complete deductive system along with the theorems of the usual first-order logic resulting therefrom (e.g. the Finiteness Theorem).

*Step 3*: Here we shall arrange for the second sort introduced in Step 2 to be a system of subsets of the first sort, by adding certain "structure axioms". We shall obtain a complete deductive system, and the corresponding theorems from first-order logic will continue to hold. In principle, we shall find ourselves actually in first-order logic. However, what need no longer hold is the condition (required in Step 1) that the system of subsets (= the second sort) be the full power set of the first sort. Structures satisfying this condition are sometimes called "standard structures"; those violating this condition will then be called "nonstandard structures".

Through this extension of the class of structures beyond the class considered in Step 1, we shall have gained something – namely, a return to first-order logic – and we shall have, on the other hand, lost something – namely, our faith in the full power set.

## ***Second-Order Monadic Logic*** **(Step 1):**

Let $L = (\lambda, \mu, K)$ be a first-order formal language, as introduced earlier in this book (1.2.0.10). We enrich the alphabet with *second-order variables*:

$$V_0, V_1, V_2, \ldots.$$

We write VBL for the set of these variables. If $\mathfrak{A}$ is an $L$-structure with universe $A$, we call a function $H : \text{VBL} \to P(A)$ an *evaluation of the second-order variables in* $\mathfrak{A}$. Here,

$$P(A) = \{B \mid B \subseteq A\}$$

is the *power set* of $A$.

We likewise enrich the set of $L$-terms with the second-order variables. We introduce the new *atomic formulae*

$$X(t),$$

where $X \in \text{VBL}$ and $t$ is a first-order term. If $\varphi$ is a formula that we have already constructed, then $\forall X\, \varphi$ will also be counted as a formula. The formulae obtained by the above recursion are called *the formulae of second-order monadic logic*. Free and bound occurrences of second-order variables are defined analogously to those of first-order variables; in particular, we write $\mathrm{FR}(\varphi)$ for the set of all second-order variables with free occurrences in $\varphi$. If $\mathrm{Fr}(\varphi) = \emptyset$ and $\mathrm{FR}(\varphi) = \emptyset$, then we call $\varphi$ a *sentence*.

Next, if $\mathfrak{A}$ is an $L$-structure, $\varphi$ is a second-order formula, and $h : \mathrm{Vbl} \to A$ and $H : \mathrm{VBL} \to P(A)$ are evaluations, then we define the *satisfaction* relation

$$\mathfrak{A} \models^{(2)} \varphi\,[h,H]$$

by recursion on the construction of $\varphi$, as follows:

$$\begin{array}{lll} \mathfrak{A} \models^{(2)} X(t)\,[h,H] & \text{if and only if} & t^{\mathfrak{A}}\,[h] \in H(X). \\ \mathfrak{A} \models^{(2)} \pi\,[h,H] & \text{if and only if} & \mathfrak{A} \models \pi\,[h], \end{array}$$

where $t$ is any first-order term, and $\pi$ is any first-order atomic formula.

$$\begin{array}{lll} \mathfrak{A} \models^{(2)} \neg\varphi\,[h,H] & \text{if and only if} & \mathfrak{A} \not\models^{(2)} \varphi\,[h,H]. \\ \mathfrak{A} \models^{(2)} (\varphi \wedge \psi)\,[h,H] & \text{if and only if} & \mathfrak{A} \models^{(2)} \varphi\,[h,H] \text{ and } \mathfrak{A} \models^{(2)} \psi\,[h,H]. \\ \mathfrak{A} \models^{(2)} \forall x\,\varphi\,[h,H] & \text{if and only if} & \mathfrak{A} \models^{(2)} \varphi\,\left[h\binom{x}{a},H\right] \text{ for all } a \in A. \\ \mathfrak{A} \models^{(2)} \forall X\,\varphi\,[h,H] & \text{if and only if} & \mathfrak{A} \models^{(2)} \varphi\,\left[h,H\binom{X}{B}\right] \text{ for all } B \in P(A). \end{array}$$

The alteration $H\binom{X}{B}$ of $H$ just above is defined by analogy with that of the first-order evaluation $h$ (1.5.0.3).

Now let $L$ be the language of ordered rings, in which we formulated, e.g. the axioms of a real closed field (1.6). Let $\Sigma^{(2)}$ be the (finite) axiom system consisting of the axioms of an ordered field (viz. $K_1$–$K_9$, $OK_1$ and $OK_2$; recall (1.6)), together with the "cut axiom":

$$\forall X,Y\,[\exists x X(x) \wedge \exists y Y(y) \wedge \forall x,y\,(X(x) \wedge Y(y) \to x \le y) \to \\ \exists z \forall x,y\,(X(x) \wedge Y(y) \to x \le z \le y)].$$

The models of $\Sigma^{(2)}$ are the ordered fields in which every Dedekind cut is realized. It is well known that, up to isomorphism, there is exactly one such field, namely, the field $\mathbb{R}$ of real numbers. (For a proof, see, e.g. [Prestel–Delzell, 2001, 1.4.1 and 1.1.5].) Therefore, in second-order monadic logic we can axiomatize $\mathbb{R}$ up to isomorphism. As a consequence, we obtain:

**Theorem B.1.** *In second-order monadic logic there can be no relation $\vdash^{(2)}$ satisfying, for any set $\Sigma$ of sentences and any formula $\varphi$:*
*(a) $\Sigma \vdash^{(2)} \varphi$ if and only if $\varphi$ holds in every model of $\Sigma$, and*
*(b) $\Sigma \vdash^{(2)} \varphi$ if and only if $\Pi \vdash^{(2)} \varphi$ for some finite subset $\Pi$ of $\Sigma$.*

*Proof*: If there were such a complete (a) and finitary (b) deducibility relation, this would immediately imply a finiteness theorem (analogous to (1.5.6)). As an application of this one would be able to prove a second-order analogue to Theorem 2.1.1: every axiom system $\Sigma$ possessing an infinite model must possess models of arbitrarily large cardinality. This would contradict the fact that $\Sigma^{(2)}$ possesses exactly one, single infinite model (viz. $\mathbb{R}$). □

## *Two-Sorted Logic* (Step 2):

Now we introduce two independent sorts of variables (of equal status). Let

$$v_0, v_1, v_2, \ldots \text{ be variables of the first sort, and}$$
$$w_0, w_1, w_2, \ldots \text{ be variables of the second sort.}$$

The letters $x, y, z$ will always denote variables of the first sort, and the letters $a, b, c$ will always denote variables of the second sort. By introducing function symbols and relation symbols for the first sort, and, separately, other such symbols for the second sort, we can introduce terms and formulae of the first sort, and correspondingly of the second sort, as usual. Then a *structure* of this language has the form

$$\mathfrak{A}_{1,2} = \langle A_1, A_2; \ldots \rangle,$$

where $A_1$ is the nonempty universe for the first sort, and $A_2$ is the nonempty universe for the second sort. The interpretation of each function and relation symbol considered so far is always a function or a relation on $A_1$ or $A_2$, as appropriate.

We can, however, also allow functions and relations *between* $A_1$ and $A_2$. Then the syntax must be correspondingly extended. Instead of carrying this out in full generality, we restrict ourselves to one example: fields $K$ with a valuation into an ordered, Abelian group $\Gamma$.

For this purpose, we recall from Section 4.3 the concept of a valuation on a field:

$$v : K \to \Gamma \cup \{\infty\}.$$

Here, $K = \langle K; +, -, \cdot; 0, 1 \rangle$ is a field, $\Gamma = \langle \Gamma; +; 0; \leq \rangle$ is an ordered Abelian group, $\infty$ is an object larger than all elements of $\Gamma$ (with $\gamma + \infty = \infty + \gamma = \infty$ and $\infty + \infty = \infty$ for all $\gamma \in \Gamma$), and $v$ is a surjective map for which the valuation axioms

$$v(x) = \infty \text{ if and only if } x = 0,$$
$$v(xy) = v(x) + v(y), \text{ and}$$
$$v(x+y) \geq \min\{v(x), v(y)\}$$

hold for all $x, y \in K$. We leave it to the reader to express these axioms in a formal language with two sorts of variables. Here, variables of the first sort run over the field elements, and those of the second sort run over the group elements together with the object $\infty$, i.e. over the universe $\Gamma_\infty := \Gamma \cup \{\infty\}$. The function symbol for $v$, when applied to a term of the first sort, produces a term of the second sort.

All theorems of first-order logic hold for such a two-sorted formal system. In particular, there is a complete (finitary) concept of deduction. This follows easily from the fact that it is possible to translate each two-sorted language into a one-sorted language. We carry this out below for the example of valued fields.

Let

$$\mathfrak{A}_{1,2} = \langle A_1, A_2; \ldots \rangle$$

be a two-sorted structure such that $A_1 \cap A_2 = \emptyset$; it may be a model of a two-sorted axiom system $\Sigma_{1,2}$. We associate with such a structure a one-sorted structure

$$\mathfrak{A}_{1\cup 2} = \langle A_1 \cup A_2; \ldots; A_1, A_2 \rangle.$$

Now $A_1$ and $A_2$ serve as additional, unary relations on the set

$$A = A_1 \cup A_2.$$

We shall soon show how to translate the set $\Sigma_{1,2}$ of two-sorted axioms into a set $\Sigma_{1\cup 2}$ of one-sorted axioms, using our example of valued fields. First, however, we must throw in the following two structure axioms:

$$\forall x\, (A_1(x) \vee A_2(x))$$
$$\forall x \neg (A_1(x) \wedge A_2(x))$$

(where we use $A_1$ and $A_2$ also as symbols for the corresponding unary relations). A syntactic problem arising during this translation will be that every function on $A_1$ or $A_2$ is no longer a function on $A = A_1 \cup A_2$ (since $A_1$ and $A_2$ are nonempty and disjoint). We shall therefore conceive of such functions as relations. Thus, we replace the valuation function $v$, for example, with a binary relation (still denoted by $v$). Then the structure axioms

$$\forall x, y\, (v(x,y) \rightarrow K(x) \wedge \Gamma_\infty(y))$$
$$\forall x\, \big(K(x) \rightarrow \exists^{=1} y\, (\Gamma_\infty(y) \wedge v(x,y))\big)$$

express that $v$ is a function from $K$ to $\Gamma_\infty$. Finally, the valuation axioms themselves now read as follows:

$$\forall x\, (v(x,\infty) \leftrightarrow x \doteq 0),$$
$$\forall x, y, z, a, b, c\, (\cdot(x,y,z) \wedge \circ(a,b,c) \wedge v(x,a) \wedge v(y,b) \rightarrow v(z,c)), \text{ and}$$
$$\forall x, y, z, a, b, c\, (+(x,y,z) \wedge v(x,a) \wedge v(y,b) \wedge v(z,c) \wedge a \leq b \rightarrow a \leq c).$$

Here, $+$ and $\cdot$ are ternary relation symbols representing the addition and multiplication on $K$, and $\circ$ is a ternary relation symbol representing the addition on $\Gamma_\infty$. Of course, for $+$, $\cdot$, and $\circ$ we must also include structure axioms corresponding to those for $v$.

All the methods and theorems of first-order logic now apply to the language for structures of the form $\mathfrak{A}_{1\cup 2}$ and the translation $\Sigma_{1\cup 2}$ of the axioms $\Sigma_{1,2}$. For example, for every infinite, valued field $v : K \to \Gamma_\infty$, there is always a countable, elementarily equivalent valued field $v' : K' \to \Gamma'_\infty$. In particular, there is also a complete (finitary) deductive system for two-sorted logic, namely, the one from first-order logic applied to the translation.

## *Reduced Second-Order Logic* (Step 3):

Now we try to understand the language of second-order monadic logic as a two-sorted language, and to translate it into first-order logic, as in Step 2 above. We must consider two-sorted structures of the form

$$\mathfrak{A}_{1,2} = \langle A_1, A_2; \ldots \rangle,$$

where, in the standard structures considered in Step 1, $A_2$ is the power set of $A_1$. If we perform the translation into first-order logic, we shall again need representing relation symbols $A_1$ and $A_2$ for the new universe

$$A = A_1 \cup A_2.$$

Moreover, we must linguistically capture the application of an element $B$ of $A_2 = P(A_1)$ to an element $a \in A_1$. Recall that $B$ is a subset of $A_1$, and $B(a)$ now means that $a$ is an element of $B$. For this, we introduce the new binary relation $\in$, and we add the new binary relation symbol $\varepsilon$ to the language.

Now the second-order axiom system $\Sigma^{(2)}$ for complete ordered fields can be translated into a first-order axiom system $\Sigma$. Applying first-order theorems, we finally obtain a countable structure

$$\mathfrak{A}'_{1,2} = \langle A'_1, A'_2; \ldots ; \mathrel{\underline{\in}} \rangle$$

that is elementarily equivalent to

$$\langle \mathbb{R}, P(\mathbb{R}); \ldots ; \in \rangle.$$

In $\mathfrak{A}'_{1,2}$, $\mathrel{\underline{\in}}$ is the interpretation of the symbol $\varepsilon$; thus, it is a relation between elements of $A'_1$ and elements of $A'_2$.

Obviously $A'_1$ is again an ordered field; it is therefore (countably) infinite. Since $A'_2$ is at most countable, it obviously cannot be the power set of $A'_1$. What, then, is $A'_2$?

As we shall soon see, $A'_2$ is actually isomorphic to a *subset* of the power set $P(A'_1)$ of $A'_1$. Since $\mathfrak{A}'_{1,2}$ satisfies the axiom system $\Sigma^{(2)}$, "cuts belonging to $A'_2$" are realized in $A'_1$. But since $A'_2$ does not contain all the cuts on $A'_1$, $A'_1$ cannot be Dedekind complete.

In order to see that one can view $A'_2$ as a subset of $P(A'_1)$, consider the function

$$g : A'_2 \to P(A'_1)$$

defined by $g(b) := \{a \in A'_1 \mid a \mathrel{\boxminus} b\}$ for $b \in A'_2$.

This function is injective, since in our standard structure $\langle \mathbb{R}, P(\mathbb{R}); \ldots; \in \rangle$, the extensionality axiom $\mathrm{ZF}_1$,

$$\forall a, b\, (\forall x\, (\varepsilon(x,a) \leftrightarrow \varepsilon(x,b)) \to a \doteq b) \tag{B.1.1}$$

(Section 1.6), holds. Since $\mathfrak{A}'_{1,2}$ is elementarily equivalent to $\langle \mathbb{R}, P(\mathbb{R}); \ldots; \in \rangle$, (B.1.1) holds also in $A'_{1,2}$. Thus, $g(b_1) = g(b_2)$ implies $b_1 = b_2$. The injectivity of $g$ means that we can identify $A'_2$ with a subset of $P(A'_1)$, as claimed.

Our example shows, in particular, that, on the one hand, all sentences about the $\in$-relation between $\mathbb{R}$ and $P(\mathbb{R})$ expressible in the two-sorted language, carry over to $A'_1$ and $A'_2$, and, on the other hand, this is never sufficient to guarantee that $A'_2$ is the "full" power set of $A'_1$.

# References

[Barwise, 1977] J. Barwise (ed.), *Handbook of Mathematical Logic*, Studies in Logic, Vol. 90, North Holland, 1977.

[Chang–Keisler, 1973–90] C.C. Chang and H.J. Keisler, *Model Theory*, Studies in Logic and the Foundations of Mathematics, Vol. 73, North-Holland, 1973, 1977, 1990.

[Engler–Prestel, 2005] A. Engler and A. Prestel, *Valued Fields*, Springer Monographs in Mathematics, 2005.

[Hermes, 1965–69] H. Hermes, *Enumerability, Decidability, Computability: An Introduction to the Theory of Recursive Functions*, Academic Press, Springer, 1965, 1969. (Translation of *Aufzählbarkeit, Entscheidbarkeit, Berechenbarkeit: Einführung in die Theorie der rekursiven Funktionen*, Grundlehren Math., Vol. 109, Springer, 1961; Heidelberger Taschenbücher, Vol. 87, Springer, 1971, 1978.)

[Hilbert–Bernays, 1934–68] D. Hilbert and P. Bernays, *Grundlagen der Mathematik*, Vol. I, Grundlehren der Mathematischen Wissenschaften in Einzeldarstellungen, Vol. 40, Springer, 1934, 1968; J.W. Edwards, Ann Arbor, Michigan, 1944.

[Hodges, 1997] W. Hodges, *A Shorter Model Theory*, Cambridge University Press, 1997.

[Levy, 1979–2002] A. Levy, *Basic Set Theory*, Springer, 1979; Dover Publications, 2002.

[Marker, 2002] D. Marker, *Model Theory: An Introduction*, Graduate Texts in Mathematics, Vol. 217, Springer, 2002.

[Poizat, 1985] B. Poizat, *Cours de théorie des modèles*, Villeurbanne, 1985.

[Prestel–Delzell, 2001] A. Prestel and C. Delzell, *Positive Polynomials: From Hilbert's 17th Problem to Real Algebra*, Springer Monographs in Mathematics, 2001.

[Prestel–Roquette, 1984] A. Prestel and P. Roquette, *Lectures on Formally p-Adic Fields*, Lecture Notes in Mathematics, Vol. 1050, Springer, 1984.

[Shelah, 1978] S. Shelah, *Classification Theory and the Number of Non-isomorphic Models*, Studies in Logic, Vol. 92, North Holland, 1978.

[Shoenfield, 1967] J.R. Shoenfield, *Mathematical Logic*, Addison-Wesley Series in Logic, 1967.

[Steinitz, 1930] E. Steinitz, *Algebraische Theorie der Körper*, W. de Gruyter, Berlin-Leipzig, 1930.

[Tent–Ziegler, 2011] K. Tent and M. Ziegler, *A Course in Model Theory*, Lecture Notes in Logic, Assoc. for Symbolic Logic, Cambridge Univ. Press, 2011 (in press).

A. Prestel, C.N. Delzell, *Mathematical Logic and Model Theory*, Universitext,
DOI 10.1007/978-1-4471-2176-3, 

# Glossary of Notation

$\mathfrak{A} \overset{\Gamma}{\rightsquigarrow} \mathfrak{B}$, 105
$\mathfrak{B}^S/\mathscr{D}$, 94
$\prod_{s\in S}\mathfrak{A}^{(s)}/\mathscr{D}$, 90
$x'$ $(= x \cup \{x\})$, 55
$z \cap x$ (intersection), *56*
$\neg\alpha$ (negation of $\alpha$), 3, 9, 15, 38
$\wedge$ (conjunction, "and"), 9, 15
$\vee$ (disjunction, "or"), 11, 16
$\bigwedge_{i=1}^{n} \varphi_i$ (finite conjunction of $\varphi_i$), 14
$\bigvee_{i=1}^{n} \varphi_i$ (finite disjunction of $\varphi_i$), 14
$\forall$ (for all), 9
$\forall v_0, v_1, \ldots, v_n\, \varphi$ (universal closure of $\varphi$), 13
$\forall$ (generalization rule), 17
$\doteq$ (equal-symbol in object language), 9
$\neq$ (negation of $\doteq$), 11
, ) ( (punctuation), 9
$\to$ ("implies"), 11
$\leftrightarrow$ ("equivalent"), 11
$\exists$ ("there exists"), 11
$-$ (operation on $\{T, F\}$), 15
$\sqcap$ (operation on $\{T, F\}$), 15
($\wedge$ B$_1$) (a derived rule), 17
($\wedge$ B$_2$), ($\vee$ B$_1$), ($\vee$ B$_2$), ($\leftrightarrow$ B$_1$), ($\forall$ B), (K$\forall$) (derived rules), 18
($\leftrightarrow$), ($\wedge$), ($\vee$) (derived rules), 19
$\vdash$ ($\Sigma \vdash \varphi$: $\varphi$ provable from $\Sigma$), 22
$\approx$ (equivalence relation on CT), 33
$\bigcup u$ (union of $u$), 55
$\cup$ (as in $x \cup y$), 55
P$(u)$ (power set of $u$), 55
$\subseteq$ (subset relation), 54
$\{x\}$ (singleton of $x$), 55
$\equiv$ (elementary equivalence), 61
$\cong$ (isomorphism of structures), 61

$\emptyset$ (empty set), 55
$\{u, v\}$ (unordered pair), 55
$\cong$ (isomorphism of structures), 65
$\tau : \mathfrak{A} \hookrightarrow \mathfrak{A}'$ (a monomorphism or an embedding), 65
$\tau : \mathfrak{A} \leftrightarrow \mathfrak{A}'$ (an isomorphism), 65

$(\mathfrak{A}, \sigma)$, 75
$\mathfrak{A} = \langle |\mathfrak{A}|;\ (R_i^{\mathfrak{A}})_{i\in I};\ (f_j^{\mathfrak{A}})_{j\in J};\ (c_k^{\mathfrak{A}})_{k\in K}\rangle$ (a structure), 37
$\mathfrak{A} \models \varphi$ ($\varphi$ holds in $\mathfrak{A}$), 40
$\mathfrak{A} \models \varphi\,[a_0, \ldots, a_n]$, 70
$\mathfrak{A} \models \varphi\,[h]$ ($\varphi$ holds in $\mathfrak{A}$ under $h$), 38
$\mathfrak{A} \overset{\exists}{\rightsquigarrow} \mathfrak{A}'$, 84
$|\mathfrak{A}|$ (universe of $\mathfrak{A}$), 37
$\underline{a}$ (for $a \in \underline{K}$), 75
(A1) & (A2) (quantifier axioms), 16
AK$_n$ (axioms for algebraic closedness), 53
$A_0, A_1, \ldots$ (sentential variables), 15

$\mathfrak{B} \preceq \mathfrak{A}$ ($\mathfrak{B}$ an elementary substructure of $\mathfrak{A}$), 71
$\mathfrak{B} \subseteq \mathfrak{A}$ ($\mathfrak{B}$ a substructure of $\mathfrak{A}$), 70

$\mathfrak{C} = \langle \mathbb{C};\ +^{\mathbb{C}},\ -^{\mathbb{C}},\ \cdot^{\mathbb{C}};\ 0^{\mathbb{C}},\ 1^{\mathbb{C}}\rangle$ (an algebraically closed field), 53
$\mathbb{C}$, *see* complex numbers
card$(A)$ (cardinality of set $A$), 62
$c_k$ (constant symbols), 9
$c_k^{\mathfrak{A}}$ (interpretation of $c_k$ in $\mathfrak{A}$), 37
C$_p$ (axiom for characteristic $p$), 53
(CP) (a derived rule), 18
CT (set of constant terms), 33

$D(\mathfrak{A})$ (diagram of $\mathfrak{A}$), 76
D$_n$ (axioms for $\mathbb{Z}$-groups), 51
Ded$(\Sigma)$ (deductive closure of $\Sigma$), 47

A. Prestel, C.N. Delzell, *Mathematical Logic and Model Theory*, Universitext,
DOI 10.1007/978-1-4471-2176-3, © Springer-Verlag London Limited 2011

# Index

A page number is *italicized* in this index
when that page contains a definition of the word being indexed.



A. Prestel, C.N. Delzell, *Mathematical Logic and Model Theory*, Universitext,
DOI 10.1007/978-1-4471-2176-3, © Springer-Verlag London Limited 2011 